OUR URBAN FUTURE

OUR URBAN FUTURE

AN ACTIVE LEARNING GUIDE TO SUSTAINABLE CITIES

SABINA SHAIKH AND EMILY TALEN

THE MIT PRESS   CAMBRIDGE, MASSACHUSETTS   LONDON, ENGLAND

© 2023 Massachusetts Institute of Technology

All rights reserved. No part of this book may be reproduced in any form by any electronic or mechanical means (including photocopying, recording, or information storage and retrieval) without permission in writing from the publisher.

The MIT Press would like to thank the anonymous peer reviewers who provided comments on drafts of this book. The generous work of academic experts is essential for establishing the authority and quality of our publications. We acknowledge with gratitude the contributions of these otherwise uncredited readers.

This book was set in Stone Serif and Stone Sans by Westchester Publishing Services. Printed and bound in the United States of America.

Library of Congress Cataloging-in-Publication Data

Names: Shaikh, Sabina, author. | Talen, Emily, 1958– author.
Title: Our urban future : an active learning guide to sustainable cities / Sabina Shaikh and Emily Talen.
Description: Cambridge, Massachusetts : The MIT Press, [2023] | Includes bibliographical references and index.
Identifiers: LCCN 2023008847 (print) | LCCN 2023008848 (ebook) | ISBN 9780262546843 (paperback) | ISBN 9780262376686 (epub) | ISBN 9780262376679 (pdf)
Subjects: LCSH: City planning. | Sustainable development.
Classification: LCC HT166 .T3425 2023 (print) | LCC HT166 (ebook) | DDC 307.1/216—dc23
    /eng/20230407
LC record available at https://lccn.loc.gov/2023008847
LC ebook record available at https://lccn.loc.gov/2023008848

10 9 8 7 6 5 4 3 2 1

CONTENTS

ACKNOWLEDGMENTS vii
INTRODUCTION ix

1 THE SUSTAINABLE CITY 1
2 URBAN ECOSYSTEM SERVICES 9
3 THE RURAL-TO-URBAN TRANSECT 25
4 GREEN SPACE 37
5 DENSITY 55
6 NEIGHBORHOOD DIVERSITY 73
7 MOBILITY 93
8 RESOURCE PLANNING IN CITIES 113
9 ENVIRONMENTAL JUSTICE 137

NOTES 147
INDEX 149

ACKNOWLEDGMENTS

We would like to extend hearty thanks to:

- The students in our Urban Design with Nature course (which we have been teaching since 2018) for their enthusiasm and fresh perspective
- The College and the Franke Institute for the Humanities at the University of Chicago for instructional support via the "Big Problems" program
- Nina Olney and Peyton Sanborn for their invaluable teaching and research assistance
- Beth Clevenger and Anthony Zannino at the MIT Press for their first-rate editorial guidance
- Our families for their steadfast love and support

INTRODUCTION

On our shelves is an abundance of books with some variant of the word "sustainable" in the title. Put "cities" or "urban" in the mix, alongside "smart" or "livable," and you need a new bookcase! But this literature is not designed for active learning in the college classroom. What is needed, and what we want this book to provide, is a way for you to gain both the theory and the concepts of urban sustainability—a "hands-on" understanding of the what, why, and how of sustainable cities.

There is so much packed into urban sustainability that it's easy to get lost. Anyone can become deflated by the complexity involved, struggling with the thought of needing to account for the interaction effects of urban and environmental "systems"—urban forms, land and ecosystems, climate and energy, human development, economic growth, housing and transportation, and community well-being.[1] Too often, the principles and goals of urban sustainability hover as platitudes, making urban sustainability seem unapproachable and overwhelming.

This book separates urban sustainability into a set of manageable topics. Our disciplinary slant is more social science and less biophysical environment, and the activities we chose are not exhaustive (we do not cover buildings, for example). But they do include what is arguably the essential subject matter of urban sustainability.

Using a menu of suggested classroom activities, we provide a practical way to engage with these topics and work through their thorniest aspects. Our goal is to ground what might otherwise sound nebulous, like the need to "respect nature" or "promote neighborhood diversity."

Many of the activities require making a proposal. We start with the identification of problems to be addressed and then work through possible interventions and what criteria to apply to determine how to proceed. Where should policy interventions be prioritized, and where would intervention have the most impact? We want to help you decide what information is needed to fully understand a particular sustainability problem and how to address it.

CHAPTERS

CHAPTER 1: THE SUSTAINABLE CITY
In our first chapter, we lay out the parameters of our subject, including some historical background and our own working definition of *urban sustainability*. We then present some theoretical underpinnings, along with the values and principles we believe are central to the topic.

CHAPTER 2: URBAN ECOSYSTEM SERVICES
Nature plays an important role in city life. But how can humans and nature coexist in the city? How can the dichotomy between conservation and development be framed as an integrated system of flows between natural and built environments? This chapter explores the use of ecosystem services as a framework for considering the benefits humans receive from nature, specifically applied to the urban environment.

CHAPTER 3: THE RURAL-TO-URBAN TRANSECT
The transect is an analytical method that organizes the elements of cities (lots, blocks, types of spaces) in ways that preserve the integrity of different kinds of environments, whether more urban or more rural. These environments vary along a continuum that ranges from less intensity—rural—to high intensity—urban. How can we determine what elements in a given place are more "rural" and what elements are more "urban"?

CHAPTER 4: GREEN SPACE
The ecological, economic, and health benefits of urban green space have long been recognized, yet challenges to allocating scarce urban space to greenery require careful planning and consideration of other urban sustainability goals. How can we measure green space, and what are the possibilities and challenges for increasing green space in cities?

CHAPTER 5: DENSITY
Urban sustainability means that cities should be compact rather than sprawling. What are the social, economic, and environmental gains associated with density, and under what conditions does density cease to offer clear benefit (e.g., when density becomes crowding)? How can cities grow inward and upward gracefully, increasing in density in ways that are welcomed rather than resisted?

CHAPTER 6: NEIGHBORHOOD DIVERSITY
A fundamental goal of urban sustainability is that neighborhoods should be socially and economically diverse—mixed in income, mixed in use, and actively supportive of places that commingle people of different races, ethnicities, genders, ages, occupations,

and households. How realistic is such a goal? What levels of mix currently exist? How can we measure existing levels of mix and suggest areas of the city where increasing social diversity makes the most sense?

CHAPTER 7: MOBILITY

Urban sustainability requires multimodal forms of transportation, especially walking, biking, and shared mobility such as public transit and bus rapid transit. What are the costs and benefits of investing in public transit and bicycle infrastructure? What areas of the city will reap the most benefit from such investments? What interventions should be made to help improve the ability of people to get around in the city without relying on a car?

CHAPTER 8: RESOURCE PLANNING IN CITIES

With nearly 4.5 billion people living on approximately 2 percent of global landmass, and projections of the world's urban population to increase from about 55 percent today to 70 percent by 2050, cities represent a great experiment in the allocation of scarce resources. Cities have put forth ambitious targets for net zero, carbon-free energy, water conservation, and waste reduction. What strategies can they employ? What type of governance is needed? How is success measured?

CHAPTER 9: ENVIRONMENTAL JUSTICE

Environmental justice has been posited as a framework, discourse, norm, value, rule, behavior, executive order, and social movement. What are the historical causes of environmental injustices? How are social conditions and built environments precursors to environmental and health disparities? How can city climate action strategies and urban planning better address environmental injustices and ensure that social equity is at the forefront of all sustainability solutions?

We hope you will use this book as a reference—a book you can consult to find essential resources and perspective in the pursuit of creating sustainable cities. And if this book adds some increment, no matter how small, of advancement toward a more sustainable human habitat—by giving you, as future leaders, the knowledge you need to make cities more just and equitable, more green and healthy, more progressive and thriving—we will have achieved our goal.

We challenge you to step out of your comfort zone and think more like a generalist than a specialist, which is, to us, a key tenet of what it means to be a practitioner of urban sustainability.

1

THE SUSTAINABLE CITY

Much is made of the fact that we are a rapidly urbanizing world, with many more people living in cities than at any other time in history. There is some fear attached to this reality—cities can seem to be headed in the wrong direction environmentally, socially, and economically. Uneven growth, decline, environmental harm, social injustice, and economic instability are just a few of the contemporary challenges that are being played out and manifested in cities around the globe. There's no denying that cities produce an exponential increase in flows of water, energy, and air—all of which strain social, economic, and environmental systems. To make matters worse, the benefits of living in cities are disproportionally applied, resulting in social disparities in education, access, and standard of living, leading to even greater social and economic costs. In short, cities pollute, exacerbate social inequalities, and often seem complicit in capitalism's project of "creative destruction."

This might seem like all gloom and doom, but there is a hopeful side, too. People everywhere are working to mitigate the harms of urban growth. They are applying innovative technical solutions that turn harmful environmental effects into assets. They are celebrating the coming together of people within urban neighborhoods and the creativity, empowerment, and sense of community that are unleashed. They are working to enable participation in the global process of stimulating bottom-up forms of economic and social exchange. They are using policy tools to ameliorate the negative effects of sprawl and exurban development on ecosystem services and biodiversity. They are designing and building compact urban forms that can lead to less energy consumption, less need for heating and cooling, and more opportunity for gray and recycled water systems.

In short, there is a lot of energy being put into the task of balancing environmental, social, and economic goals in pursuit of green, just, and economically viable cities—aka, the sustainable city. Our task—the task of students of urban sustainability—is to be mindful of the problems before us, learn how to recognize and measure them, know how to gauge uneven and harmful impacts, and develop the skills, tools, and policy responses needed to help redress harm and inequity. Cities *can* simultaneously provide equitable access to resources, minimize environmental degradation, and stimulate economic opportunity—but we need to develop a collective understanding of what that entails.

The purpose of this chapter is to present the conceptual basis of the "sustainable city." We present our own working definition and lay out what we believe the associated values and principles are. We start by offering some historical context. How did we arrive at this notion of the "sustainable city"?

FROM ENVIRONMENTALISM TO SUSTAINABILITY

Environmentalism gained momentum in the years following the publication of Rachel Carson's *Silent Spring* (1962), which documented the devastating effect of human carelessness on animal species. In the US, game-changing environmental acts followed: the Clean Air Act (1963), the National Environmental Policy Act (1969), the Clean Water Act (1972), and the Endangered Species Act (1973). Environmental legislation tended to focus on singular topics rather than the interactions among environmental factors, which later became the core of sustainability.

Some observed that the environmentalism that emerged in the 1970s seemed to have a kind of embedded human–nature duality to it: humanity was viewed as profane, and nature was viewed as sacred. Historian William Cronon (1996) argued that the "ideological underpinning" of the environmentalist movement was "wilderness" and that "wilderness" is a problematic concept if it is viewed as something wholly separate from ourselves. The problem is that this separation gives license to remain aloof from our everyday inhabited world. Cronon (1996) stated, "By imagining that our true home is in the wilderness, we forgive ourselves the homes we actually inhabit" (p. 81). Such views possess the untenable paradox that the only way to save nature is for humans—and the cities they live in—to not be part of it.

Most environmentalists no longer hold this kind of rigid dualistic view pitting cities against nature. The sister field of ecology has been especially helpful in working through the conceptual logjam, creating new ways of thinking about how human and natural worlds can be conceptualized as interdependent. There is now wide recognition that human behavior must be incorporated into urban ecosystem models and that much can be gained by studying cities as urban ecosystems (Melosi, 1999). Integration is achieved by lessening the environmental impacts of human design, typically via an appropriate choice of materials, renewable energy sources, and a keen sensitivity to ecological context.

People realized that ecology could provide some useful, heuristic analogies for the study of human environments—and the development of proposals for more environmentally sound urban development. For example, Jane Jacobs (1961) argued persuasively that the diversity of a healthy city is analogous to the diversity of a natural ecosystem. Urban economists promoted the same idea by arguing that dense, diverse cities breed innovation and that the resultant knowledge accumulation and spillover effects are a vital component of economic growth (Sassen, 1994). If cities separated uses into functional zones—a current condition of metropolitan form—they digressed significantly from natural systems in which interdependencies create and maintain a healthy diversity.

By the 1980s, the concept of "sustainability" was defining the imperative of healthy human–environment interaction. Although "sustainability" is a word that has been used for centuries, modern usage of the term began with the 1987 publication *Our Common Future* by the Brundtland Commission—the UN Commission on Environment and Development. The report popularized the concept of "sustainable development," which it defined as development that "meets the needs of the present without compromising the ability of future generations to meet their own needs."

There was an obvious alignment between sustainability theory and the human–nature integration that environmentalists and ecologists had been working through. Sustainability was the balancing of economic, environmental, and social needs, which is another way of framing the idea that it is necessary to find a balance between human-made and natural environments. What emerged was a new brand of environmental thinking: cities no longer viewed as necessarily detrimental but as part of the solution to environmental problems.

Activity 1.1: Sustainability Goals (Concept Map)

The United Nations Development Programme (UNDP) adopted a set of seventeen sustainability goals in 2015. Working in teams, review the goals and create a hierarchical concept map. The main topic—sustainability—should be placed at the center of the map, and the other goals should be ranked, ordered, and clustered. Think about the interrelations among goals. Are some connected to others? Are some goals on their own?

DEFINING THE SUSTAINABLE CITY

Many years after the publication of *Our Common Future* in 1987, there are now many definitions of sustainable cities and urban sustainability (we treat these terms synonymously). Drawing on these definitions, we settled on our own working definition that we use for this book:

The sustainable city is a city that exerts minimal damage on the environment while maintaining a diverse and resilient economy, where the distribution of positives (resources) is equitable and the distribution of negatives (noxious uses) does not burden neighborhoods unfairly, and where local residents have a sense of community and a genuine stake in decision-making.

The idea is that if cities have these qualities, they can be sustained in the long run. To be sustainable means to last, to endure—and in the context of urban sustainability, that means that cities, towns, and other human habitats need to reduce their consumption of resources, increase their resilience and adaptability, and strive for social equity at every turn.

> The New Urban Agenda was adopted at the United Nations Conference on Housing and Sustainable Urban Development (Habitat III) in Quito, Ecuador, on 20 October 2016. It was endorsed by the United Nations General Assembly and adopted by consensus by all 193 countries of the United Nations. It is intended to represent "a shared vision for a better and more sustainable future." It has been critiqued as being too vague, too focused on development as a solution (and thus too "neoliberal"), and too lacking in implementation specifics.

Note that terms like "sustainable city" and "urban sustainability" imply a long-term goal and a vision of a desired future. Terms like "sustainable development" and "sustainable urbanization," on the other hand, refer to the many pathways and processes involved in achieving the sustainable city. We believe that both are essential: the sustainable city is more than just green buildings and pervious pavement; it involves the process of designing walkable communities along with connections to transit, food, green space, health care, and amenities. It means minimizing the harms of flooding, heat, hazards, noxious uses, and other environmental risks. It means that underlying all of these goals is a commitment to social equity and meaningful community engagement. It is the vision of desired ends and the processes and pathways for getting there.

This definition means that it's essential to include a wide range of topics—subject matter that is usually not combined in a typical university course. It's important to absorb all sides of sustainability—the environmental side, the economic side, the social side, and the design side. A serious student of urban sustainability does not get to ignore the sides of sustainability they don't feel comfortable with. They might prefer one side of what sustainability means, but in a learning environment, they need to be exposed to the full range of thinking. In fact, that's the challenge of urban sustainability. It is by definition multidisciplinary and therefore complex. Urban sustainability requires constant assessment, and assessment requires consideration of trade-offs, cross-purposes, and cause–effect dynamics. Classrooms are ideal places to consider all of these dimensions.

VALUES AND PRINCIPLES

Values and principles are intrinsic to sustainability. What counts as more versus less sustainable, the kinds of metrics that are used, and even the sustainability topics that are selected to define basic concepts are all intrinsically normative—driven by an understanding of "what ought to be." For example, the priorities of access and equity, the valuing of diversity over homogeneity, the imperative of transit and bicycle-based mobility over car-based transportation, and in general the need to acknowledge the fundamental importance of reducing energy consumption, minimizing waste, and protecting habitat—all of these principles have a normative basis.

Selected values and principles are also guided by theories about the nature of human–environment interaction. Earlier theories about this interaction ranged from determinism (that the environment shapes human culture) to possibilism (that human–environment interaction is conditioned by cultural practices). An ecosystem services approach focuses on quantifying the ways that humans directly benefit from the environment. Others conceptualize human–environment interaction in terms of dependence on the environment or modification and adaptation of the environment.

For the purposes of this workbook, we adopt a normative, action-oriented approach. This means that although we are interested in describing human–environment interaction, our focus is on learning how to react in ways that are simultaneously positive for the environment and positive for humans, all in the context of cities.

This is not meant to be overly prescriptive, and as we have stressed, urban sustainability is also about managing the processes of change; process and ideals—principles that involve a normative understanding of urban sustainability goals—are not in opposition. Creating visually explicit models of future development *and* providing an inclusive process for getting there are both needed to help resolve the human-versus-nature duality problem.

This translates to a need to be specific and avoid platitudes. We might know what the sustainable city is in broad terms, but it is an ongoing task to define parameters, offset perverse effects, and figure out ways to more effectively implement the essential objectives. The definition of the sustainable city is, in principle, resolved; the question is how to get there and how to stay on target. Should our strategy involve trying to find a better process that guides city building toward a more sustainable outcome or via a stronger articulation of what the sustainable city is supposed to be? Undoubtedly, it is both.

There is no single approach or model that can be universally applied. What is needed are tools and strategies that can help us make intelligent decisions. There is usually never one right answer—there are alternative answers that prioritize different aspects of urban sustainability. Solutions must be found, but they have to be supported with facts and well-structured reasoning.

Activity 1.2: Sustainability Data and What It Reveals (Team Research)

Students will work in teams to explore sustainability data available for their city or for an assigned city. This could be obtained from a city's sustainability plan or from a city's data portal. Many cities have data online through government portals—in Chicago, for example, data is found at the Chicago Data Portal. Teams will explore what data is available with the goal of creating a visualization of sustainability metrics. An example might be a map of Leadership in Energy and Environmental Design (LEED)–certified buildings, parks, or green roofs throughout the city. Where are these located in the city? Teams might also compare their map of green resources or other sustainability metrics to a map of income, education, race, or ethnicity. Are there associations to be made?

PUTTING VALUES INTO PRACTICE

Normative values and principles do not translate easily into practice—the crafting of policies, programs, and prescribed urban patterns and forms. Principles are one thing; prescribing actual interventions is another. In this book, we encourage you to tackle this by weighing options, fully understanding the trade-offs involved, and knowing how to craft a well-reasoned argument in support of a given proposal.

In the quest for sustainable cities, there are value judgments and trade-offs to be made. It is often not possible to maximize sustainability across all environmental goals—perhaps an increase in access to transit means building more density, and perhaps that density interferes with protection of a riparian corridor. But that doesn't mean that urban sustainability is indeterminable—rather, it means that the translation from value and principle to implementation and practice requires explicit recognition of the trade-offs and priorities being made.

A key value concerns social equity. So much of what urban sustainability is about is fairness: who are the winners and losers of sustainability outcomes? Who gets to make the decisions, and how do others get to have a voice and a stake in the decision-making process? Who is harmed by an unsustainable practice, and who most benefits in the attempt to build more sustainably? How has history shaped disparities related to unsustainable conditions, and how can we change that?

In urban sustainability, justice and equity play out in specific and measurable ways—what is important is the recognition that these impacts can and should be determined. What populations disproportionately suffer from the failure to build cities in sustainable ways? Who is most burdened by strains on limited resources? Which populations are experiencing more than their fair share of pollution exposure, flooding, drought, and heat waves? The other side of the coin is the distribution of the *benefits* of sustainable cities—things like walkability, access to transit, and green space development. But here again there are winners and losers—that is, the "gains" are unevenly distributed.

These justice and equity topics are central to all of the topics we include in this workbook—not just in terms of understanding the level of injustice and inequity involved, but in terms of understanding how one might go about proposing alternative solutions.

Activity 1.3: Sustainability Plan Evaluation (Homework)

Read through your city's sustainability plan and provide an evaluation. Is the plan clear and concise? Is there attention paid to necessary trade-offs involved in achieving sustainability? Does the plan identify priorities, or are all goals weighted equally? Does the plan lay out a realistic way of achieving its stated goals? What is the plan's approach to achieving social justice and equity?

Students can make use of sustainability rating systems as a guide to their evaluation. Two sustainable cities rating systems that might be useful are U.S. Green Building Council's (USGBC's) Leadership in Energy and Environmental Design for Neighborhood Development (LEED ND) and Ecodistricts. Both are focused specifically on the neighborhood plan. In what ways does your city's sustainability plan guide future neighborhood development? Do you think rating systems are valuable for understanding sustainability, and are they useful for understanding the value of your city's sustainability plan?

LITERATURE CITED

Carson, R. (1962). *Silent spring*. Houghton Mifflin.

Cronon, W. (1996). The trouble with wilderness; or, getting back to the wrong nature. In *Uncommon ground* (pp. 69–90). W. W. Norton.

Jacobs, J. (1961). *The death and life of great American cities*. Vintage.

Melosi, M. V. (1999). *The sanitary city: Urban infrastructure in America from colonial times to the present*. Johns Hopkins University Press.

Sassen, S. (1994). *Cities in a world economy*. Pine Forge Press.

2

URBAN ECOSYSTEM SERVICES

Nature plays a critical role in healthy and vibrant cities. The dynamic relationship between built and "natural" is manifested in outcomes related to environmental quality, human and ecological health, and social conditions. But what is nature? How is it defined? Are humans part of nature or an external force that needs find a sustainable coexistence in the city? On a practical level, land is scarce, even fixed in some cases, leading to necessary trade-offs between conservation and development. But human well-being is critically dependent on healthy ecosystems. So, how can this conventionally considered dichotomy between conservation and development be reframed as an integrated and synergistic system of flows between natural and built environments? The perspective of ecosystem services is often used to consider and measure the trade-offs and synergies that exist between green space protection and built environment, conservation and economic growth, and private and public provision of nature. This chapter explores the use of ecosystem services as a framework for considering the specific benefits humans receive from nature, specifically applied to the urban environment, and assesses the pros and cons of such an approach. This leads into the next chapter, which considers the benefits and costs associated with urban green infrastructure investments using an ecosystem service framework.

INTRODUCTION

RESOURCES

Daily, G. C. (1997). Introduction: what are ecosystem services? In G. Daily (Ed.), *Nature's services: Societal dependence on natural ecosystems* (pp. 1–10). Island Press.

Gomez-Baggethun, E., de Groot, R., Lomas, P. L., & Montes C. (2010). The history of ecosystem services in economic theory and practice: From early notions to markets and payment schemes. *Ecological Economics, 69*(6), 1209–1218.

Millennium Ecosystem Assessment (MEA). (2003). *Ecosystems and human well-being: A framework for assessment.* Island Press. (Summary)

Ecosystem services have been defined broadly as the benefits humans receive from nature. Specifically, Daily (1997) defines ecosystem services as "the conditions and processes through which natural ecosystems, and the species that make them up, sustain and fulfill human life." Inherently an anthropocentric concept, ecosystem services have been motivated by a need to better consider environmental outcomes in decision-making processes, by framing nature as a type of capital. Early models of natural resource economics had always implicitly considered nature as capital, in the form of an input to the production of final goods and services. The modern ecosystem service framing builds on developments in the field of environmental economics, which explicitly recognizes the value of nature in a protected state to one that acknowledges the controversies of the commodification of nature's benefits (Gómez-Baggethun et al. 2010). This framing of nature as capital is intended to ensure its inclusion as a component of society's productive base, along with physical or manufactured capital, human capital, and social capital. While human life is dependent on ecosystem services, humans have long been altering the capability of ecosystems to provide such services (MEA, 2003). These connections between humans and ecosystems are bidirectional, interconnected, scale dependent, dynamic, and critically important to understand as we chart toward a sustainable urban future.

DISCUSSION QUESTIONS

1. What is the Millennium Ecosystem Assessment? What are its purpose and goals?
2. What are the connections between ecosystems and human well-being? How do humans contribute to ecosystem degradation, and how, in turn, are humans affected by degraded ecosystems?
3. What are the pros and cons of an approach that frames nature as capital?

Activity 2.1: Cities and Nature (Concept Map)

Respond to the following two questions using a concept map:

1. What is the role of nature in cities?
2. What is the role of cities in nature?

TOPIC 1: CONCEPTUAL FRAMEWORK FOR ECOSYSTEM SERVICES

RESOURCES

de Groot, R. S., Wilson, M. A., & Boumans, R. M. J. (2002). A typology for the classification, description and valuation of ecosystem functions, goods and services. *Ecological Economics, 41*(3), 393–408.

Provisioning services	Cultural services	Regulating services
Food	Recreation	Climate regulation
Fresh water	Aesthetic	Water regulation and filtration
Wood for fuel	Spiritual and religious	Waste decomposition
Fiber	Cultural heritage	Pollination
Energy	Educational	Disturbance control
Medicine	Sense of place	
	Community cohesion	

Supporting services
Nutrient cycling
Soil formation
Primary production
Biodiversity

2.1 Ecosystem functions and services. Regulating and supporting ecosystem services provide indirect benefits to humans, while provisioning and cultural ecosystem services provide direct benefits to humans.

Millennium Ecosystem Assessment (MEA). (2003). *Ecosystems and human well-being: A framework for assessment.* Island Press. (Chapters 1–4)

Ecosystem services have often been categorized and classified by the type of services they provide (MEA, 2003). Figure 2.1 shows a commonly used classification that features Provisioning Services, which directly provide goods and services to humans like food and fiber; Cultural Services in the form of recreation and education; Regulating Services such as climate regulation, water filtration, disturbance and storm prevention, and pollination, which benefit humans through indirect pathways; and Supporting Services, which contribute to the provision of all other ecosystem services (de Groot et al. 2002; MEA, 2003).

Based on the categorization of the types of services, figure 2.2 from de Groot et al. (2002) outlines the complete process of ecosystem service assessment, the steps of which will be discussed throughout this chapter. Using this framework, the *total value* of ecosystem services comprises human values based on economics and sociocultural values and preferences, as well as ecological values, which are often converted into human terms through monetization in order to aid decision-making. Aside from regulating services, figure 2.2 outlines an alternative naming of ecosystem functions to figure 2.1, where habitat function refers to the generation and maintenance of wild plant and animal habitat and biodiversity, production function aligns with Provisioning Services, and information function represents Cultural Services. It is important to note that ecosystem services are not independent of each other, nor are they separable. There is significant overlap, time and scale dependence, and joint

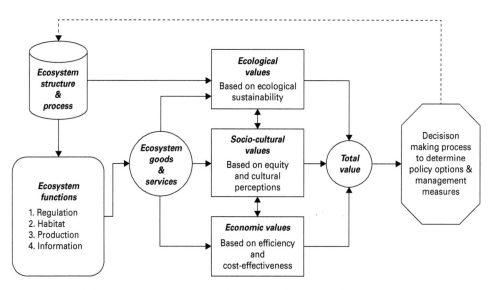

2.2 Framework for integrated assessment and valuation of ecosystem functions, goods, and services. *Source*: Reprinted from de Groot, R. S., Wilson, M. A., & Boumans, R. M. J. (2002). A typology for the classification, description and valuation of ecosystem functions, goods and services. *Ecological Economics, 41*(3), 393–408.

production of ecosystem services, which is necessary to accurately represent natural systems but presents challenges in decision-making.

> **Activity 2.2:** Classification of Ecosystem Services (Jigsaw)
>
> With a partner or group, consider one of the following categories in figure 2.1 of the classification of ecosystem services: Provisioning, Cultural, Regulating, and Supporting. Give some examples of ecosystem services that would fit in this specific category. What type of natural system would provide such ecosystem services? What is the relationship of these ecosystem services to people? For example, how might people benefit from these services? And, in turn, how might people enable the provision of these services?
>
> Then, come together with your classmates. Look for any overlap in your identification of ecosystem services or relationships across categories. How do people fit in across categories?

TOPIC 2: MEASUREMENT AND METHODS FOR ECOSYSTEM SERVICES

RESOURCES

Goulder, L. H., & Kennedy, D. (1997). Interpreting and estimating the value of ecosystem services. In P. Kareiva, H. Tallis, T. H. Ricketts, G. C. Daily, & S. Polasky (Eds.), *Natural capital: Theory and practice of mapping ecosystem services* (pp. 15–33). Oxford University Press.

Millennium Ecosystem Assessment (MEA). (2003). *Ecosystems and human well-being: A framework for assessment*. Island Press. (Chapter 6)

Ecosystem services have been widely valued and monetized in the economic literature. Studies have provided dollar-value estimates of ecological units like trees (McPherson et al. 1997) to entire green spaces in cities (Kremer et al. 2016) to global urban agriculture (Clinton et al. 2018) and even the entire world's ecosystem services (Costanza et al. 1997). To economists, a utilitarian framework implies that ecosystem services offer value directly (food), indirectly (storm prevention, recreation, culture), and simply from existing. All types of value are considered to create aggregate measures of total value, as an attempt to consider all direct and indirect paths from ecosystems to human well-being.

This conceptual notion that we can value something for its existence, regardless of tangible interaction, is known as "nonuse value" or "existence value." This is based on the idea that humans derive utility or satisfaction from knowing something exists, like animal welfare, or from an option that may lead to future value, like a medicinal discovery. Existence value may also be derived from spiritual or cultural associations with nature. While existence value is widely acknowledged, the tools for its measurement remain controversial.

For ecosystem services that yield direct or indirect use value, however, methods for quantification of the monetary value of ecosystem services range from cost estimates to more sophisticated approaches of economic valuation. *Avoided* costs are often used as a way to monetize the benefits of green infrastructure and ecosystem services. These include reductions in energy or water bills, reduced maintenance costs, lower health expenditures, or avoided damage costs associated with flooding or other hazards. *Replacement* costs have also been applied as the cost to replace a lost ecosystem service. For example, if natural stormwater mitigation is lost to development, a stormwater fee may be assessed based on having to supply that service through either a gray infrastructural system or preservation of natural stormwater nearby.

Neither avoided cost or replacement cost approaches utilize what economists refer to as "economic welfare." This notion, based on microeconomic theory, focuses on value rather than price (i.e., cost), which can best be thought of as how much something is worth above and beyond its "sticker" price. This is particularly important with respect to nature and human health, which have no directly observable market price despite the many attempts to commodify them. For example, the cost of an asthma condition due to pollution may be measured by the price of medication. However, the full welfare value of this cost would include lost productivity, pain and suffering, and loss of leisure activity. Economic valuation methods that rely on welfare analysis attempt to get at this value by observing actual or stated behavioral responses to changes in environmental quality to determine the "willingness to pay" for improvements. The methods include the use of *hedonic pricing*, which assesses the incremental property value associated with proximity, access, or exposure to parks; the *travel cost method*, which focuses on recreation values from the ability to visit parks or other amenities; and *stated preference methods*, the most contentious of the methods, which rely on individuals' stated willingness to pay for a hypothetical improvement in environmental quality or natural resource accessibility. These stated preference methods have been used mainly to measure nonuse and existence values where no interaction with nature can be easily observed.

Another method increasingly used with respect to climate resilience considers insurance premiums and payouts. While this type of data is not always publicly available, and concerns exist about using the monetized values of insurance as indicators of welfare (Green et al. 2016), there is valuable behavioral information to be gained in understanding what people spend to protect themselves against shocks, hazards, and chronic stressors in specific regions, cities, or neighborhoods.

Other ecosystem service values may be less suited for economic valuation techniques and may require more qualitative or observational types of methods. The existence of social and cultural benefits afforded by ecosystem services is widely accepted, but the measurement may require a more intentional consideration in overall value assessment. For a detailed overview of ecosystem service valuation and its use in urban planning, see Gómez-Baggethun and Barton (2013).

DISCUSSION QUESTIONS

1. How do humans "reveal" their preferences and values for ecosystem services?
2. What is "existence value"? Can you value something you never see, touch, or interact with directly or indirectly? How might someone you know value it?

Box 2.1
Discount Rates

An important aspect of the valuation of ecosystem services is the consideration of the time horizon and how to value benefits that occur in the future, particularly the far-off future. In any investment, costs may occur upfront and be ongoing while benefits accrue over time. In the case of nature, there are longer time horizons than in traditional private financial investments or real estate due to the slow, long pace of ecological processes. In addition, investments in nature can appreciate over long time horizons (e.g., tree growth), whereas investments in physical infrastructure necessarily depreciate or disappear over long time spans. The "discount rate" is the conventionally used mechanism to compare future values to present ones by discounting the value of money received later compared to money spent today. It is quite valid to discount future returns as most people would prefer a dollar today than to receive that same dollar 1, 10, or 100 years from now. Even adjusted for inflation, this is likely to hold true because the dollar invested today will yield more than that dollar in the future, as determined by prevailing interest rates. For private investments, positive discount rates of around 7 percent, based on the returns from average alternative investments in stocks or housing, are commonly used, while 3 percent or so is more appropriate for a rate of return in public investments. A contentious debate has emerged over the appropriate discount rate to use for environment and nature, particularly climate change, which unlike private investments has an intergenerational time horizon. This debate emerges from the premise that we can't observe individual trade-offs made over such a long time span, and therefore, there is little information revealed to infer the appropriate discount rate. This debate is manifested in the calculations for the social cost of carbon for climate change–related decision-making, in which the choice of a discount rate has contributed in part to widely varying estimates of

URBAN ECOSYSTEM SERVICES

> **Box 2.1**
> continued
>
> the damages associated with greenhouse gas emissions from fossil fuels. In practice, this has significant consequences as it affects the rate of return of ecosystem service projects, which necessarily require longer time horizons for full realization.

Activity 2.3: Net Present Value of Ecosystem Investments (Homework)

The formula for discounting is the inverse of compounding, which you might use to calculate how your savings account balance grows over time. With discounting, we are interested in *present value*—the calculation of the value of a future return today. The formula to be used for a one-time payoff in the future is:

Present value = *future value* / $(1+r)^t$,

where r is the discount rate, and t is the year the future value occurs. As in the case for all investments, benefits and costs will accrue over time, and this formula extends to one of *net present value (NPV)*, which aggregates discounted benefits (*B*) minus costs (*C*) over the full time horizon of the project (*T*).

$$\text{Net present value } (NPV) = \sum_{t=1}^{T} (B_t - C_t)/(1+r)^t$$

One example of urban ecosystem services are those provided by trees. Many cities have proposed large-scale tree-planting projects (e.g., MillionTreesNYC) as a way to create urban habitat, mitigate air pollution and urban heat, manage water, and sequester carbon, among other functions.

Use the above formula to consider a tree-planting project that generates ecosystem services over different time horizons. For simplicity, say your city plans to plant 1 million trees over the next five years. The tree-planting project is expected to cost $100 million annually in years 1 to 5, while the benefits (tree ecosystem services) would begin after trees mature enough starting in year 10. The benefits are estimated to be $25 million annually from year 10 until the end of the project time horizon (T).

Problem: Use the information and the formula provided to make the NPV calculations to fill in table 2.1.

Table 2.1 NPV of the Million Tree Planting Project

	Time Horizon (T)			
Discount rate (r)	5	10	50	100
0 percent				
3 percent				
7 percent				
10 percent				

DISCUSSION QUESTIONS

1. Describe some of the project benefits—the ecosystem services from trees.
2. Propose a method for monetizing these ecosystem services from trees.
3. Using a strict decision rule that requires NPV > 0, the project may or may not be adopted depending on the discount rates and time horizon selected. Why are the NPV estimates so different? What do you think the appropriate discount rate should be for ecosystem service investments like this one? Why would a 0 percent discount rate be hard to justify in practice?
4. What are some potential sources of uncertainty associated with your estimates?

Activity 2.4: Monetization of Nature? (Debate)

Biological, philosophical, and economic debates have emerged to address the appropriateness of monetizing nature. Attempts to place a value on the world's nature (Costanza et al. 1997) have been widely criticized by noneconomists and economists alike for different reasons. Some opponents of valuing ecosystem services question the morality of putting a dollar value on nature or human lives, while other opponents (economists) who are not against monetizing nature, per se, argue that current attempts to do so make simplified assumptions, grossly underestimate nature's value and serve only to create awareness and controversy (Toman, 1998).

Nevertheless, it is still argued that pricing nature can help protect it by ensuring its value is accounted for in decision-making. Policies are often prioritized through explicit or implicit comparisons of costs and benefits, and environmental ones are no exception. If decision makers allocated funding and investments to conservation, it inherently becomes a monetary exercise. Without proper accounting of economic value, land in conservation uses will be deprioritized relative to land in development uses for which financial values readily exist. While traditional cost-benefit analysis and resulting policy and land use decisions may have largely ignored ecological values, methods for valuation and considerations for the appropriateness of use have evolved greatly since the 1970s. An ecosystem service framework, while still reliant on human values for nature, highlights the value of environmental protection and conservation to decision makers.

Students will consider these questions in relation to ecosystem service valuation. The class will be divided into teams to debate the following proposition: *Dollar values should not be placed on human lives or nature, and decisions involving environment and natural resources should be made without monetization.*

TOPIC 3: URBAN ECOSYSTEM SERVICES

RESOURCES

Bolund, P., & Hunhammar, S. (1999). Ecosystem services in urban areas. *Ecological Economics, 29*(2), 293–301.

Gómez-Baggethun, E., & Barton, D. N. (2013). Classifying and valuing ecosystem services for urban planning. *Ecological Economics, 86*, 235–245.

Table 2.2 Examples of urban ecosystem services, adapted from Gómez-Baggethun and Barton (2013) and McPhearson et al. (2014)

Urban Feature	Ecosystem Services	Benefit
Community garden	Food supply, stormwater retention, social space	Food consumption/sales, reduced flooding, social connections
Trees	Heat mitigation, carbon storage and sequestration, air purification, water storage, habitat, aesthetics	Reduced energy use, reduced flooding, recreation experiences, improved health, views
Habitat	Animal sightings, recreation, pollination services, species biodiversity	Recreation experiences, existence values, views
Vegetation	Storm barriers, air purification, noise reduction, water regulation and runoff mitigation	Improved health, reduced property and personal damage
Green roofs	Heat mitigation, carbon sequestration, stormwater retention and runoff mitigation	Reduced energy use, improved health, views, productivity
Parks	Recreation and cognitive development, social space	Improved physical and mental health, social connections

Sources: Gómez-Baggethun, E., & Barton, D. N. (2013) and McPhearson, T., Hamstead, Z. A., & Kremer, P. (2014).

McPhearson, T., Hamstead, Z. A., & Kremer, P. (2014). Urban ecosystem services for resilience planning and management in New York City. *Ambio, 43*(4), 502–515.

Millennium Ecosystem Assessment (MEA). (2003). *Ecosystems and human well-being: A framework for assessment*. Island Press. (Chapter 5: "Dealing with Scale")

Ecosystem services have evolved in part as a tool for measuring and achieving success related to United Nations frameworks, including the Sustainable Development Goals, Millennium Development Goals, and Convention on Biological Diversity. The approach has been applied to global conservation of biodiversity (Turner et al. 2007), large land areas like watersheds, and whole ecosystems like wetlands and forests. Growing recognition of the usefulness of the framework to support decision-making and urban planning, combined with an increased focus on city resilience, led to more urban applications of ecosystem services. These approaches were not immediately transferable. The complex and fragmented socioenvironmental landscapes in cities required updated frameworks and tools specific to *urban* ecosystem services to consider the novel human–nature interactions that occur in urban environments and to better inform urban planning and design (Bolund & Hunhammar, 1999; Gómez-Baggethun & Barton, 2013; Jansson, 2013; McPhearson et al. 2014).

Urban ecosystem services are generated via a variety of sources in cities. Table 2.2 shows examples of six types of common urban green infrastructure and the associated ecosystem services.

The assessment of ecosystem services is a multistage process, which includes identification, measurement, valuation, and provision. It is also a multiscalar problem, and the scales and boundaries of an assessment are neither arbitrary nor politically neutral (MEA, 2003). Ecosystem service provision occurs at site-specific, neighborhood, city, regional, national, and even global scales. Note the range of scales present in table 2.2, where the features are regional or local and even site based in some cases, while the associated ecosystem services can extend to global scales, as with the case of carbon sequestration. In this case, cities can be active sites of ecosystem service provision that reach large populations of people, despite the fragmented landscapes and inelasticity of land supply.

DISCUSSION QUESTIONS

1. Why are cities, with large, dense populations of humans, important places for ecosystem service provision?
2. What kind of ecosystem services can be provided in cities? How?
3. Is there a trade-off between achieving population density and ensuring space for ecosystem service provision in cities?

Box 2.2
Space to Grow Green Schoolyards in Chicago

Space to Grow is Chicago's green schoolyards program, led by managing partners The Healthy Schools Campaign and Openlands. The green schoolyards engage Chicago Public School populations and community residents in a design process to transform school grounds into healthy, green spaces within the built environment for students and residents to play and learn. The elements of green schoolyards can include outdoor classrooms, play equipment, nature areas, and gardens while also attending to community-identified needs like localized flood control through state-of-the-art rainwater collection and storage design. In addition, the green schoolyards are part of the community, both designed through intensive community participation and input, and open to community regardless of school affiliation. As such, the program aims to provide an array of benefits to children and communities related to engagement and cohesion, health and wellness, education and learning, and environment (Space to Grow, 2022). Indeed, research has shown that green schoolyards can reduce stress and increase resilience in children and enhance positive youth development outcomes related to physical activity and prosocial behavior (Bates et al. 2018; Chawla et al. 2014). In addition, co-benefits of the schoolyard design include the provision of ecosystem services, locally through recreation and social gathering spaces, flood control, and heat mitigation; regionally from water management and air quality control; and globally through carbon sequestration and greenhouse gas mitigation.

URBAN ECOSYSTEM SERVICES

Box 2.2
continued

2.3 Space to Grow green schoolyard at Grissom Elementary School in the Hegewisch community on the southeast side of Chicago, before and after transformation. *Source*: Space to Grow website: https://www.spacetogrowchicago.org/.

TOPIC 4: ECOSYSTEM SERVICES IN PLANNING AND PRACTICE

RESOURCES

de Groot, R. S., Alkemade, R., Braat, L., Hein, L., & Willemen, L. (2010). Challenges in integrating the concept of ecosystem services and values in landscape planning, management and decision making, *Ecological Complexity, 7*(3), 260–272.

Fisher, B., Turner, R. K., & Morling, P. (2009). Defining and classifying ecosystem services for decision making. *Ecological Economics, 68*(3), 643–653.

Hansen R., Frantzeskaki, N., McPhearson, T., Rall, E., Kabisch, N., Kaczorowska, A., Kain, J., Artmann, M., & Pauleit, S. (2015). The uptake of the ecosystem services concept in planning discourses of European and American cities. *Ecosystem Services, 12*, 228–247.

Jansson, Å. (2013). Reaching for a sustainable, resilient urban future using the lens of ecosystem services. *Ecological Economics, 86*, 285–291.

Millennium Ecosystem Assessment (MEA). (2003). *Ecosystems and human well-being: A framework for assessment.* Island Press. (Chapter 8)

Questions left unanswered so far include *who* will provide ecosystem services, *why* will they provide ecosystem services, and *to whom* will they provide the services? Under the current economic and political system, it is unlikely that sustainable cities can emerge through public investment alone. Will the estimated values of ecosystem services be sufficient to facilitate comparisons of investment in urban green infrastructure versus other land uses? How will public and private landowners differ in their response to information on the value of green investments? While there is no simple answer, the recognition of value does not always translate to the realization of value. This is in part due to the diffuse, public nature of ecosystem services and lack of market prices associated with the benefits. For example, the US Environmental Protection Agency (EPA) currently sets the social cost of carbon at $51 per ton of carbon dioxide, which makes green infrastructure that sequesters carbon "valuable," yet it remains unclear how any city or individual investing in such carbon mitigation would capture such a value. This is where policy mechanisms and economic incentives can play a significant role. In the economic sense, if value exists, it can be captured through a market or market-based approach such as a fee or payment. Environmental fees act to discourage actions that degrade nature and include stormwater fees, impact fees, fines for improper disposal of environmental hazards, road tolls, gas taxes, or other punitive measures. These are based on a framework of *negative externalities*, where one's self-interested actions do not take into account the costs imposed on others, in the form of pollution or environmental degradation.

Another market-based framework relies on payments, sometimes termed "payments for ecosystem services." These are based on a conceptual framing of *positive externalities*, where one's actions result in environmental benefits to others. There have been programmatic attempts to motivate such actions, including tax breaks for hybrid or electric vehicles, green roof grants, rebates for energy-efficient appliances, rain barrel giveaways, or subsidies for clean energy. In theory, these types of programs would generate enough return on investment to be self-sustaining over time if the benefits could be captured appropriately.

A significant controversy arises in relying on market values for environmental protection. If one has to essentially "pay" for better environmental conditions, there will necessarily be inequitable outcomes. Some would argue that this is the case for any commodity in a market-based economy, but should the environment be treated as any other commodity, and should health and access be based on income?

DISCUSSION QUESTIONS

1. How is ecosystem services provision dependent on externalities and incentives? Does it need to be? Why not just mandate ecosystem services be provided by everyone living in cities? Why not just provide solely through public investment?
2. How much can individuals affect ecosystem service provision? Is an individual-level framing the appropriate framing for large, intractable global problems like climate change?

Activity 2.5: Program Design for Ecosystem Services (Homework)

Propose: Consider a way to increase the provision of ecosystem services in your neighborhood (campus, town, city). It could be something as local as encouraging the reduction of food waste at your school or more broadly local like providing your city's business owners with payments or tax rebates to convert lawns or pavement to native plant gardens.

Design a program that creates an incentive for people to provide these ecosystem services, noting the behavioral trade-offs, potential obstacles or pitfalls, and the overall sustainability and longevity of the program. Your program should address the following:

1. Describe your idea
2. Motivate your program. Why is it important? What will it accomplish?
3. What ecosystem services does it provide, and why are these valuable? Why would a behavioral nudge be an appropriate way to induce provision?
4. Why do you think it could be achieved through behavioral mechanisms? What else would have to happen to make the behavioral nudge effective (e.g., infrastructure, regulation, education)? Be specific.
5. What challenges or obstacles do you foresee, and how might you address these?

Write this up as a proposal to whomever would need to be convinced given the scale you've proposed (e.g., your school, a homeowners association, business owners, the mayor or city council) and not as a list of answers to the preceding questions. Use images, tables, or other tools to make it appealing to the reader. Cite literature as needed to use appropriate formatting to list references.

Ecosystem services have increasingly entered the urban planning domain (Gómez-Baggethun & Barton, 2013; Jackson, 2003; McPhearson et al. 2014). The understanding of the services provided by nature strengthens the case for green space integration into city design made by Burnham and others in the nineteenth century. Still, while the ecosystem services framework poses a promising tool for incorporating green infrastructure into planning, policy, and decision-making, challenges limit effective implementation.

Box 2.3
Challenges to Using Ecosystem Services in Practice

1. **Classification.** While much progress has been made in defining ecosystem services, standardization is needed (de Groot et al. 2002; Fisher et al. 2009). The interconnectedness and dynamic nature of ecological processes, as well as the distinction between intermediate and final services, complicate the ability to separate one service from another (e.g., wildlife habitat and recreation). This "joint production" of ecosystem services can lead to problems with identification and double counting of services and must be carefully considered.
2. **Timing.** Investments in nature accrue and even appreciate over time. As discussed earlier, discounting plays a major role in assessing investments over time. Investments in nature or green infrastructure necessarily require longer time horizons, which using the practice of discounting become valued less in the future. Planners may need to consider longer time horizons with lower discount rates to compare green infrastructure projects to gray ones.
3. **Space.** In addition to physical and social connectivity, ecological connectivity is an important concept when considering green infrastructure. The location of distributed green space matters due to localized environmental conditions like heat, flooding, air quality, and built environment characteristics such as proximity to highways, landfills, and industrial facilities. Further, the benefits associated with ecosystem services are directly dependent on the number and concentration of people affected by the services (Kozak et al. 2011). To fully capture the benefits provided by nature, strategies related to urban form and development must be informed by an understanding of the connection of people to water flows and hydrology, wildlife habitat and migration routes, air currents, and other dynamic and spatially connected systems. One such approach might be to incorporate ecological connectivity into urban land-use zoning.
4. **Scale.** Ecosystem service provision can come at a site scale like a single building (green roof) or yard (tree, habitat) and at large or even global scales (watersheds, forests). While a site decision can be made by a building manager or property owner, watershed management may involve multiple jurisdictions and government entities. Further, the ecosystem services provided and associated benefits to humans may accrue at different scales. For example, an urban forest may be managed by a city government, yet the benefits associated with carbon sequestration may accrue to a global population. This mismatch between the scale of production and level of management, regulation, or planning presents practical complications that require innovative partnerships and cooperation (Gómez-Baggethun & Barton, 2013; McPhearson et al. 2014). In addition, urban ecosystem services are generally provided where there is available space or where new development occurs. As a result, the supply of such services is not always aligned with where demand is the greatest. As a result, there can be significant mismatches between where ecosystem services can be supplied and where there is demand, leading to gaps in provision for lower-income people or communities and, in some cases, exacerbation of environmental injustices (Herreros-Cantis & McPhearson, 2021).
5. **Equity.** Disinvested areas suffer from a lack of high-functioning ecosystems and access to nature. Minority and ethnic communities often experience the worst environmental

> **Box 2.3**
> continued
>
> exposures and a lack of access to environmental amenities, resulting in wide environmental health disparities within cities. These outcomes have been induced by historically segregationist housing policies, past urban planning and urban form culture, systemic racism, and gentrification. Equity needs to be a core principle of urban and environmental planning, both to reduce environmental health disparities and to counter the effects of gentrification as communities become greener. The question remains whether an ecosystem service framework can effectively incorporate necessary information about community socioeconomic conditions while supporting stakeholder engagement and participatory processes. A related concern about using an ecosystem service framework lies in its monetization for planning purposes. Once nature is monetized, it becomes commodified and inevitably affordable to a few or many. In a payments-for-ecosystem services framework, only those with the ability to provide ecosystem services will be compensated, thereby creating a twofold problem in that ecosystem services will end up only in areas where people can afford to provide them, and only those who provide ecosystem services will be part of a constructed environmental value chain.

Activity 2.6: Ecosystem Services in Practice (Debate)

Proposition: Ecosystem services, while a valuable conceptual framework for understanding how humans benefit from nature, are too complicated to be applied in practice. There are too many challenges associated with (a) measuring ecosystem services; (b) monetizing ecosystem services; (c) dealing with timing, scale, and equity considerations; and (d) finding space for ecosystem services in cities where we need dense and diverse human populations. In this case, we are better off investing in ecosystem service provision along the far end of a *transect* and not near the central core of cities.

Counterresponse: Nature not only improves the quality of social and physical life in cities, but it can also make them more resilient to acute shocks and chronic stressors. With increased heat, flooding, and extreme weather, it is imperative for cities to invest in all kinds of smart, modern, and green infrastructure. This approach can save cities money and provide returns on investment to residents through reduced energy costs and water savings and to cities from increased property value bases and tourism revenues. Significant advances have been made in connecting ecosystem services to urban planning over the past two decades and will continue as cities take the lead in climate mitigation and adaptation. Cities will need to invest in an equitable distribution of green infrastructure to ensure the most vulnerable communities are protected from environmental hazards, exposures, and resulting gentrification.

LITERATURE CITED

Bates, C. R., Bohnert, A. M., & Gerstein, D. E. (2018). Green schoolyards in low-income urban neighborhoods: Natural spaces for positive youth development outcomes. *Frontiers in Psychology*, *9*(805), 1–10.

Chawla, L., Keena, K., Pevec, I., & Stanley, E. (2014). Green schoolyards as havens from stress and resources for resilience in childhood and adolescence. *Health & Place*, *28*, 1–13.

Clinton, N., Stuhlmacher, M., Miles, A., Uludere Aragon, N., Wagner, M., Georgescu, M., Herwig, C., & Gong, P. (2018). A global geospatial ecosystem services estimate of urban agriculture. *Earth's Future*, *6*(1), 40–60.

Costanza, R., d'Arge, R., De Groot, R., Farber, S., Grasso, M., Hannon, B., Limburg, K., Naeem, S., O'Neill, R., Paruelo, R., Raskin, R., Sutton, P., & Van Den Belt, M. (1997). The value of the world's ecosystem services and natural capital. *Nature*, *387*(6630), 253–260.

Daily, G. C. (1997). Introduction: What are ecosystem services? In G. Daily (Ed.), *Nature's services: Societal dependence on natural ecosystems* (pp. 1–10). Island Press.

Fisher, B., Turner, R. K., & Morling, P. (2009). Defining and classifying ecosystem services for decision making. *Ecological Economics*, *68*(3), 643–653. https://10.1016/j.ecolecon.2008.09.014.

Gómez-Baggethun, E., & Barton, D. N. (2013). Classifying and valuing ecosystem services for urban planning. *Ecological Economics*, *86*, 235–245.

Gómez-Baggethun, E., de Groot, R., Lomas, P. L., & Montes C. (2010). The history of ecosystem services in economic theory and practice: From early notions to markets and payment schemes. *Ecological Economics*, *69*(6), 1209–1218.

Green, T. L., Kronenberg, J., Andersson, E., Elmqvist, T., & Gomez-Baggethun, E. (2016). Insurance value of green infrastructure in and around cities. *Ecosystems*, *19*, 1051–1063.

Herreros-Cantis, P., & McPhearson, T. (2021). Mapping supply of and demand for ecosystem services to assess environmental justice in New York City. *Ecological Applications*, *31*(6), 1–21.

Jackson, L. (2003). The relationship of urban design to human health and condition. *Landscape and Urban Planning*, *64*(4), 191–200.

Jansson, Å. (2013). Reaching for a sustainable, resilient urban future using the lens of ecosystem services. *Ecological Economics*, *86*, 285–291.

Kozak, J., Lant, C., Shaikh, S., & Wang, G. (2011). The geography of ecosystem service value: The case of the Des Plaines and Cache River wetlands, Illinois. *Applied Geography*, *31*(1), 303–311.

Kremer, P., Hamstead, Z. A., & McPhearson, T. (2016). The value of urban ecosystem services in New York City: A spatially explicit multicriteria analysis of landscape scale valuation scenarios. *Environmental Science & Policy*, *62*, 57–68.

McPhearson, T., Hamstead, Z. A., & Kremer, P. (2014). Urban ecosystem services for resilience planning and management in New York City. *Ambio*, *43*(4), 502–515.

McPherson, E. G., Nowak, D., Heisler, G., Grimmond, S., Souch, C., Grant, R., & Rowntree, R. (1997). Quantifying urban forest structure, function, and value: The Chicago Urban Forest Climate Project. *Urban Ecosystems*, *1*, 49–61.

Millennium Ecosystem Assessment (MEA). (2003). *Ecosystems and human well-being: A framework for assessment*. Island Press.

Space to Grow. (2022). Greening Chicago's schoolyards. https://www.spacetogrowchicago.org/.

Toman, M. (1998). Why not to calculate the value of the world's ecosystem services and natural capital. *Ecological Economics*, *25*(1), 57–60.

Turner, W. R., Brandon, K., Brooks, T. M., Costanza, R., da Fonseca, G. A. B., & Portela, R. (2007). Global conservation of biodiversity and ecosystem services. *BioScience*, *57*(10), 868–873.

3

THE RURAL-TO-URBAN TRANSECT

How should urban and natural worlds be integrated? The "rural-to-urban transect" is a conceptual and analytical tool used to think about integration in practical terms. We can use it to define, debate, and propose the most appropriate integration of elements such as streets, lots, buildings, and public spaces.[1]

INTRODUCTION

RESOURCES

Falk, B. and Duany, A. (2021).Transect urbanism: Readings in human ecology. Oro Editions.

Jacobs, Jane. (1961). Chap. 19, "Visual Order: Its Limits and Possibilities." in The Death and Life of Great American Cities. NY: Vintage. Pp. 372–391.

Urban and suburban sprawl is unsustainable because it wastes land, destroys habitat, necessitates car dependence, and makes nonautomotive forms of travel like transit difficult, if not impossible. In short, people and land uses are spread around inefficiently, and ultimately, this inefficiency leads to disinvestment in the urban core.

But what is sprawl, exactly? One definition is that sprawl is the inappropriate placement of urban elements in rural contexts. For example, when large-format retail (the "big-box store") is placed in a suburban area, it can be viewed as the inappropriate placement of an urban level of commerce in a rural environment—driven by a failure to accept the limits of suburban life. In other words, the thinking goes, if people want to live close to "nature," they should accept the limits of what that means. The failure to do so is what creates sprawl.

One theory, known as "the rural-to-urban transect" ("the transect" for short), maintains that the solution to this inefficiency and waste is to build in a way that is true to a place's physical character. Rural areas should be rural and urban areas should be urban. The problem with current building—what makes it unsustainable—is that there is an inappropriate mixing of things: rural elements are found in urban places, while urban elements are found in rural places. In regard to suburban sprawl, transect theory maintains that it is necessary to limit the "urbanizing of the rural," such as office towers in otherwise pristine environments. In the early and mid-twentieth century, modernist architects like Le Corbusier thought that high-rises set in parks would not only house the masses but would clearly demarcate town from country. Yet planners were already arguing at the time that this was a "debased" approach to town development: "Rural influences neutralize the town. Urban influences neutralize the country. In a few years all will be neutrality" (Sharp, 1932, p. 11).

Conceptualizing the proper integration of town and country, or city and nature, is a topic that has engaged writers for centuries. There has been much rhetoric about the need to view the "humanmade world" as "natural," but most often cities are viewed in striking contrast to the natural world. Integrating city and nature "on the ground" has proven to be extremely difficult: witness the explosion of suburbia as an attempt to be "close" to nature.

Sometimes the resolution of urban and nature integration is accomplished by surrounding the city with a greenbelt, but the transect takes a different approach. Rather than stopping urban growth with physical barriers that underscore urban versus rural division, the goal is to connect and integrate the two realms along a continuum. The transect's approach is to accommodate a diverse range of development types at multiple points along this gradient. This has the benefit of accommodating a greater range of development choices—what are known as "immersive" environments. Immersive environments that are more rural might consist of wide streets and open swales, while immersive environments that are more urban might consist of formal boulevards and public squares.

A transect is simply a geographical cross section of a region that can be used to reveal a sequence of environments. For human environments, this cross section can be used to identify a set of habitats that vary by their level and intensity of urbanism, a continuum that ranges from rural to urban. This range of environments is the basis for organizing the components of the built world: building, lot, land use, street, and all of the other physical elements of the human habitat. The resulting urban pattern is based, theoretically, on finding the proper balance between natural and humanmade environments along the rural-to-urban transect. This balance entails an urban pattern that is sustainable, coherent in design, and composed of an array of livable, humane environments that satisfy a range of human needs. Building places that are true to their urban or rural character is the basis of sustainability.

DISCUSSION QUESTIONS

1. What example can you give from our own experience where an element—a building, a street, a public space—seems out of place: a rural element in an urban place or an urban element in a rural place?

2. What is your definition of "suburban sprawl"? Do you agree that it might be defined as an ineffective attempt to urbanize the rural?

> **Activity 3.1:** Build a Transect (Team Research)
>
> Consult the research page of the Center for Applied Transect Studies (CATS) website (www.transect.org) and investigate the different ways in which transects have been portrayed. Some transects fall under the category "pop culture," and a few examples are shown. There have also been transects of shoes, dogs, hairstyles, beverages, and cars. Students will divide into teams, and each team will be assigned a given element found in the urban environment, such as a bench, a pathway, a car, or a monument. Teams should then devise a rural-to-urban continuum for the element and find a representative image. Teams should present their transects to the class and compare results. Be prepared to justify why a given element is more versus less "urban" or "rural."

> **Activity 3.2:** Greenbelts (Homework)
>
> One of the common reactions to resolving the tension between "urban" and "nature" is to surround a town with an impenetrable greenbelt, a practice that is much more common in Europe. In the United States, greenbelts have been difficult and expensive to put into place, although two states, Oregon and Colorado, have had some success. One criticism is that development that occurs outside of the greenbelt can be just as objectionable as any of the worst sprawl the greenbelt was designed to eliminate. Using Google Earth, observe the greenbelts of cities in Colorado and Oregon. Do you find evidence that greenbelts are effectively containing sprawl? Do you think greenbelts could be used more effectively in the United States as a way to block suburban sprawl and create a healthier rural–urban balance? Write a short summary (max. 500 words) of your greenbelt exploration, and include images to illustrate your findings.

TOPIC 1: HUMAN HABITATS

RESOURCES

Johnson, S. (2001). *Emergence: The connected lives of ants, brains, cities, and software.* Scribner.

Transect theory applies ecological principles to the design of cities by thinking of cities as a range of habitats. Scientists have observed that nature conforms to a certain spatial ordering of ecosystems, a progression of biodiversity that ranges from prairie to woodland or tundra to foothill. The transect mirrors this principle, hoping to apply this principle to create a range of human habitats composed of varying degrees of urban intensity.

A fundamental ecological principle is that within a specified area, there exists an interrelatedness—a functional linkage—between organisms and their physical environment. In nature, each element of a community serves a constructive or at least stabilizing role. In the early twentieth century, Patrick Geddes, an influential regional

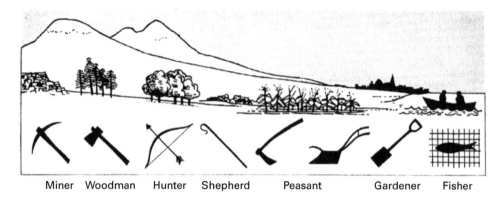

3.1 The Valley Section. *Source*: Geddes, P. (1915). *Cities in evolution*. Williams & Norgate. Public domain.

planner and polymath, proposed an idea similar to the transect: the "valley section" (figure 3.1). He argued that the valley section could be used as a device to find "the rhythms of the land masses of the earth . . . from snow to sea, from highland to lowland." By studying a place in this way, Geddes believed that each valley section had a different level of natural intensity and that these would determine what occupation was likely to be found there: miners and hunters in the higher elevations and shepherds on the grassy slopes, for example. Environment and occupation in turn determined the essential character of a place.

The transect uses a similar principle to define the appropriateness of certain types of elements within a given human habitat. Thus, elements with a lower level of urban intensity—say, a single-family house—rightly belong in less urban areas, while elements with a higher level of urban intensity—a high-rise building—belong in more urban areas. Different types of elements should be placed in a way that is appropriate to their sphere or range of influence. By design, a road with high-speed geometrics should be placed in an area that is meant to serve or connect a wide geographic range, while a narrow residential street responds to a much smaller range.

Appropriate intensity and character sound burdensome, but they are not particularly difficult to perceive. A farmhouse would not be expected and therefore would not contribute to the immersive quality of an urban core. A high-rise apartment building would not be expected, nor would it contribute to the immersive quality of a rural environment.

Of course, in natural ecosystems, the distinction between "organism" and "physical environment" may be more obvious than in humanmade environments. A street, a building, and a type of use are all elements of the built world (the "organisms," by analogy), but they also comprise the physical environment. What is important is the basic ecological principle of interrelatedness, the multidimensional nature of ecosystems in which elements are bound to context and cannot be treated as isolated,

> **Box 3.1**
> Transect Walk
>
> An urban analytical method known as a "transect walk" is a good way to understand the range of rural-to-urban habitats found in a given city. The idea is to draw lines–*transects*—across a map and then walk these routes to investigate the variety of human habitats found along them. Grady Clay was an author who popularized a similar method, walking a number of cross sections in "rigorous pursuit of generalizations along a linear path" (Clay, 1980, 1994). In Clay's "Will the Real Portland Please Stand Up?" (1998), he lists thirteen cross-section criteria, including "go where the flow begins" and "go for the center" in an attempt to find "elements that make the place tick." Interestingly, he connects his transect method not only to Patrick Geddes's "Valley Section" but also to the sixteenth-century anatomist Vesalius.

singular dimensions. Under the transect system, planning the built environment is focused on creating immersive environments comprising elements that are, in an ecological sense, interrelated.

Each habitat requires a certain degree of internal diversity. In natural ecologies, different kinds of habitats exhibit a different mix of elements, satisfying a range of different species, a theme also explored by Steven Johnson in his book *Emergence: The Connected Lives of Ants, Brains, Cities, and Software*. This idea, which ecologists refer to as "complexity," can be applied to human environments, whereby different human habitats—immersive environments—present the full range of elements necessary to create sustainable and resilient places. Transect theory maintains that it is only through this diversity of elements at the neighborhood scale that sustainable cities are achieved.

DISCUSSION QUESTIONS

1. Are you comfortable with drawing analogies between human and natural habitats? Is it necessary for human habitats to reflect ecological principles?
2. Do you think there is something intrinsically wrong with a high-rise apartment building sitting in a cornfield?

> **Activity 3.3:** Human Habitat Disruptions (Case Study)
>
> Some have observed that the principle of interrelatedness means that introducing organisms that are not adjusted to a given habitat can destabilize an entire community. This effect has been observed in urban environments as well. For example, the redevelopment of an urban core, in which large "mega projects" like stadiums are imposed on an otherwise fine-grained urban fabric, can have negative effects. Discuss the case of a large stadium built in the downtown of a major city. Good examples include Indianapolis, St. Louis, and Phoenix.

> Each team will be assigned to a stadium project as a basis for analyzing human habitat disruption. Teams should use a case study analytical approach to consider the following:
>
> - How much of the area was physically and socially altered? See if you can find before-and-after pictures of the project site to determine the level of disruption involved.
> - Is the stadium viewed as a success and a public benefit? Who are the winners and losers in this kind of single-purpose, large-format development?
> - What could have been done differently, without large-scale capital investment, to stimulate economic revitalization downtown?

TOPIC 2: TRANSECT ZONES

RESOURCES

Falk, B. and Duany, A. (2021).Transect urbanism: Readings in human ecology. Oro Editions.

The transect idea has increasingly been applied to the planning and regulation of cities, especially in the United States—specifically to zoning codes. To facilitate the application of transect principles to zoning, many local governments have adopted a standard set of six "transect zones," adapting and expanding them as needed (a full accounting of the number of transect-based zoning codes that have been proposed or adopted is available on the website www.placemakers.com). The intended result is a zoning map comprising transect zones. A diagram of these model zones is shown in figure 3.2, and their characteristics are listed in figure 3.3.

Transect zones at the rural end of the spectrum, T-1 and T-2, delineate lands that should either be preserved in perpetuity or reserved for future protection (i.e., as land becomes available). Land in these zones should not be developed. Figure 3.3 describes the buildable zones, T-3 through T-6. The T-3 zone is buildable but is meant to be the most naturalistic, least dense, most residential habitat of a community; buildings consist of single-family, detached houses; office and retail, on a restricted basis, are also permitted; buildings are a maximum of two stories; and open space is rural in character. The T-4 zone is the generalized but primarily residential habitat of a community; buildings

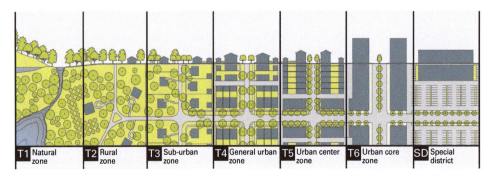

3.2 Model transect zones. *Source*: Duany, A., Sorlien, S., & Wright, W. (2008). *SmartCode: Version 9.2.* www.smartcodecentral.com

T1

T-1 NATURAL
T-1 Natural Zone consists of lands approximating or reverting to a wilderness condition, including lands unsuitable for settlement due to topography, hydrology or vegetation.

General Character:	Natural landscape with some agricultural use
Building Placement:	Not applicable
Frontage Types:	Not applicable
Typical Building Height:	Not applicable
Type of Civic Space:	Parks, Greenways

T2

T-2 RURAL
T-2 Rural Zone consists of sparsely settled lands in open or cultivated states. These include woodland, agricultural land, grassland, and irrigable desert. Typical buildings are farmhouses, agricultural buildings, cabins, and villas.

General Character:	Primarily agricultural with woodland & wetland and scattered buildings
Building Placement:	Variable Setbacks
Frontage Types:	Not applicable
Typical Building Height:	1- to 2-Story
Type of Civic Space:	Parks, Greenways

T3

T-3 SUB-URBAN
T-3 Sub-Urban Zone consists of low density residential areas, adjacent to higher zones that some mixed use. Home occupations and outbuildings are allowed. Planting is naturalistic and setbacks are relatively deep. Blocks may be large and the roads irregular to accommodate natural conditions.

General Character:	Lawns, and landscaped yards surrounding detached single-family houses; pedestrians occasionally
Building Placement:	Large and variable front and side yard Setbacks
Frontage Types:	Porches, fences, naturalistic tree planting
Typical Building Height:	1- to 2-Story with some 3-Story
Type of Civic Space:	Parks, Greenways

T4

T-4 GENERAL URBAN
T-4 General Urban Zone consists of a mixed use but primarily residential urban fabric. It may have a wide range of building types: single, sideyard, and rowhouses. Setbacks and landscaping are variable. Streets with curbs and sidewalks define medium-sized blocks.

General Character:	Mix of Houses, Townhouses & small Apartment buildings, with scattered Commercial activity; balance between landscape and buildings; presence of pedestrians
Building Placement:	Shallow to medium front and side yard Setbacks
Frontage Types:	Porches, fences, Dooryards
Typical Building Height:	2- to 3-Story with a few taller Mixed Use buildings
Type of Civic Space:	Squares, Greens

T5

T-5 URBAN CENTER
T-5 Urban Center Zone consists of higher density mixed use building that accommodate etail, offices, rowhouses and apartments. It has a tight network of streets, with wide sidewalks, steady street tree planting and buildings set close to the sidewalks.

General Character:	Shops mixed with Townhouses, larger Apartment houses, Offices, workplace, and Civic buildings; predominantly attached buildings; trees within the public right-of-way; substantial pedestrian activit
Building Placement:	Shallow Setbacks or none; buildings oriented to street defining a street wall
Frontage Types:	Stoops, Shopfronts, Galleries
Typical Building Height:	3- to 5-Story with some variation
Type of Civic Space:	Parks, Plazas and Squares, median landscaping

T6

T-6 URBAN CORE
T-6 Urban Core Zone consists of the highest density and height, with the greatest variety of uses, and civic buildings of regional importance. It may have larger blocks; streets have steady street tree planting and buildings are set close to wide sidewalks. Typically only large towns and cities have an Urban Core Zone.

General Character:	Medium to high-Density Mixed Use buildings, entertainment, Civic and cultural uses. Attached buildings forming a continuous street wall; trees within the public right-of-way; highest pedestrian and transit activity
Building Placement:	Shallow Setbacks or none; buildings oriented to street, defining a street wall
Frontage Types:	Stoops, Dooryards, Forecourts, Shopfronts, Galleries, and Arcades
Typical Building Height:	4-plus Story with a few shorter buildings
Type of Civic Space:	Parks, Plazas and Squares; median landscaping

3.3 Transect zone characteristics. *Source*: Duany, A., Sorlien, S., & Wright, W. (2008). *SmartCode: Version 9.2*. www.smartcodecentral.com

consist of single-family, detached houses and rowhouses on small- and medium-sized lots; limited office and lodging are permitted; retail is confined to designated lots, typically at corners; buildings are a maximum of three stories; and open space consists of greens and squares.

The T-5 zone is the denser, fully mixed-use habitat of a community; buildings include rowhouses, flexhouses, apartment houses, and offices above shops; office, retail, and lodging are permitted; buildings are a maximum of five stories; and open space consists of squares and plazas. T-6 is the densest residential, business, cultural, and entertainment concentration of a region; buildings include rowhouses, apartment houses, office buildings, and department stores; buildings are disposed on a wide range of lot sizes; surface parking lots are not permitted on frontages; and open space consists of squares and plazas.

Each transect zone has a certain kind of physical character, which is determined by, among other variables, building height and placement, frontage type, and civic space design. Defining zones by these kinds of physical qualities is a very different approach from the usual way of defining zones in conventional zoning codes—determined by regulating land use rather than building form.

Applying an urban-to-rural transect as a way of integrating cities and nature requires first defining what transect environments might look like. It requires differentiating, for example, between T-3 suburban, T-4 general urban, and T-5 urban center areas. Every city and town has its own set of rural-to-urban environments. Some of these environments are immersive in that the elements within them conform to a given level of intensity. Many other parts of a city do not seem to conform to any particular transect environment, consisting of mixtures of urban and rural elements.

Box 3.2
Transect Zones

How does one develop an understanding of what transect (T) zones are supposed to be like? A good place to start is to consult photographic examples. An excellent resource is Sandy Sorlien's collection (see companion website). The photographs provide examples of what T zones look like in various cities around the globe.

You can also review some generic standards for T zones. The metrics listed in figure 3.3 are part of the "Smartcode" standards, an open-source template for transect-based zoning. It is important to note that the descriptions in figure 3.3 are for comparison purposes only. All standards need to be locally calibrated, because different locales have different immersive qualities and building traditions. Chicago's T-3 suburban quality is different from Tucson's T-3 suburban quality. The question is, as you move left and right along the rural-to-urban transect, living more "urban" versus more "rural," how do the qualities of these different environments change?

DISCUSSION QUESTIONS

1. Dividing the urban landscape into six zone types under the transect approach is much simpler than conventional zoning, where there can be a hundred or more zoning categories. Do you see an advantage to keeping things simple?
2. Do you think land uses should be mixed within all types of zones, whether more urban or more rural? What would be the value of limiting the range of uses within a zone?

Activity 3.4: Transect Violation (Class Discussion)

Transect theory revolves around the idea that some elements fit better in urban locations, while other elements fit better in rural locations. In the case of housing, for example, some people believe that although neighborhoods should have a mix of housing types, there are limits. For example, single-family houses placed next to apartment buildings might undermine a block's immersive urban quality. Discuss the pros and cons of "transect violation" as a concept. Should there be a limit when it comes to the mixing of building types, and if so, what should such limits be based on? When is the mix of building types healthy and vibrant, and when is it chaotic or harmful?

TOPIC 3: NATURE IN THE CITY

RESOURCES

Steiner, F. (Ed.). (2016). *Nature and cities: The ecological imperative in urban design and planning.* Lincoln Institute of Land Policy.

Cronon, W. (1996). The trouble with wilderness; or, getting back to the wrong nature. In *Uncommon ground* (pp. 69–90). W. W. Norton & Company.

Lang, S., & Rothenberg, J. (2017). Neoliberal urbanism, public space, and the greening of the growth machine: New York City's High Line park. *Environment and Planning A: Economy and Space, 49*(8), 1743–1761.

Urban dwellers crave nature in the city, but what form should "nature" take? Are there aspects of nature that, when inserted in urban places, are ultimately disruptive? An argument that transect proponents sometimes make is that, in places at the urban end of the transect spectrum, there is a danger that social and economic connections will be undermined if insertions of unnecessarily large or misplaced "green" space occur. A few notable urbanists have made the case that this is uniquely an Anglo-American problem and that in other cultural contexts, such as in Paris, insertions of green space are much more likely to be small and nondisruptive (Kunstler, 2001).

The necessity of urban connectivity and density means it isn't always possible to daylight streams and preserve riparian corridors in urban places. There is no denying that city growth has resulted in the loss of thousands of square miles of natural areas, consumed by dwellings, stores, and offices. But should this be remedied by insertions of green space? Such interventions may clip the street grid, reducing transportation capacity and thereby density. If greening urbanism subverts walkability, then it lowers environmental performance by fostering vehicular traffic and dispersing a population, with residents adding their lower-density carbon footprints to the global crisis.

Transect theory stipulates that the response to natural conditions should be different depending on how rural or how urban a particular location is. In more rural areas, green infrastructure should be given priority. At the opposite end of the transect continuum—the urban core—urban qualities are given priority. In the urban transect zones, the primary consideration of development is to strengthen the urban fabric, which may mean that natural features are subjected to an urban treatment. If viewed in broad terms, this should not be seen as environmental insensitivity but rather as a system that ultimately ensures the preservation of natural resources.

DISCUSSION QUESTIONS

1. Frank Lloyd Wright's dream for a utopian world was that everyone would live on a one-acre plot in their own house. He called it "Broadacre City." Do you think this arrangement would give everyone access to "nature"?
2. Do you think there is a limit to how much "green space" a city should have? Is there validity to the criticism that too much green space blocks connectivity and spreads things out?

Activity 3.5: Times Square: Is It Green? (Debate)

The argument has been made that the very high-density neighborhoods of New York City are the most "green" places on earth, despite the fact that there might not be a lot of actual green space. The value of high density is not only that it maximizes social connection, but that it keeps human settlement compact and contained, minimizing its wasteful spread over the land. Further, natural features, if present, should be subjected to an urban treatment, which should not be construed as environmental insensitivity: on the contrary, concrete cities of high-rise apartments ultimately ensure the preservation of natural resources. Small parks and aligned street trees help condense the human habitat.

But others view dense urban living with its acres of asphalt and concrete as bad for people and bad for the environment. The proper human habitat, they argue, is one with ample green space. This offers multiple ecological benefits, such as rainwater capture to reduce runoff, habitat for urbanized wildlife, and reduction of the urban heat island. Having green spaces in cities helps people connect with nature, providing a higher quality of life.

Students will use Times Square in New York City as a representative example of high-intensity urbanism. The class will be divided into two teams to debate the following proposition: *Times Square in New York City is the most "green" place on earth.*

Box 3.3
The Synoptic Survey

One approach to measuring transect qualities is called the "Synoptic Survey." The Synoptic Survey is typically used in environmental analysis to determine the characteristics of a given site by discovering the habitats (or "communities") that it contains: a wetland here, an oak hammock there, a rocky outcrop there. The objective is to determine the values of each habitat in order to recommend the degree of protection and type of restoration each might require. Every functioning habitat is a symbiotic community of micro-climate, minerals, humidity, flora, and fauna.

The Synoptic Survey takes the most representative locales and analyzes them using two analytical tools: Dissect and Quadrat analysis. The Dissect analyzes the conditions above and below ground and involves borings to determine things like soil condition and the depth of the water table. The Quadrat involves taking a representative area (say 100 × 100 feet) within which the component elements of flora and fauna are itemized and counted.

Dissect and Quadrat analysis can be extended into urbanized areas (figure 3.4). Quadrat analysis involves counting or averaging elements within a given area, say, the number of housing units or the average street width. Dissect analysis entails taking a slice through the environment and measuring its characteristics, such as building height and frontage type.

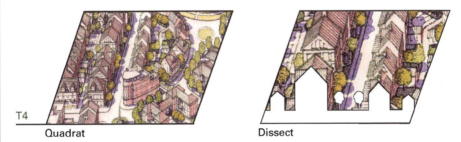

Quadrat   Dissect

3.4 Dissect and quadrat analyses. *Source*: Duany, A., Sorlien, S., & Wright, W. (2008). *SmartCode: Version 9.2*. www.smartcodecentral.com

Activity 3.6: Public Frontages (Homework)

Go to the SmartCode website, www.smartcodecentral.com, and download a pdf of the SmartCode manual version 9.2. Find table 4A, "Public Frontages." Walk down the main commercial street in your community—which will probably consist of T4, T5, or T6 zones—and photograph the frontages. Compare the frontages to the specifications in table 4A and produce a map that identifies the lots that have frontages that are not permitted, according to the manual, in the T4, T5, or T6 zones. Would it make much of a difference to pedestrian life, commercial vitality, or the overall of quality of urbanism if the frontages were different? Think about what changes you would make.

LITERATURE CITED

Clay, G. (1998). The national observer: Will the Real Portland please stand up? *Landscape Architecture*, *88*(5), 156–155. http://www.jstor.org/stable/44680373.

Clay, G. (1980). *Closeup: How to read the American city*. University of Chicago Press.

Clay, G. (1994). *Real places: An unconventional guide to America's generic landscape.* University of Chicago Press.

Kunstler, J. H. (2001). London: Landscape as the cure for cities. In *The city in mind* (pp. 225–252). The Free Press.

Sharp, T. (1932). *Town and countryside: Some aspects of urban and rural development.* Oxford University Press.

4

GREEN SPACE

Any city dweller can attest to the respite offered by neighborhood and city parks—from small pocket parks for native gardens to neighborhood gathering spots for recreation and children's activities to large city parks designed as showcases to a city's vibrant culture and life. With many of these green spaces come ecological, economic, and health benefits, but challenges to allocating scarce urban space to greenery remain. Planners must consider creating designs that accommodate both aesthetic and recreational needs, while ensuring sustainable and equitable uses of land. In addition, financing and governance of green space is complex, and the question of how green space is built, including by whom and for whom, remains contentious. This chapter considers the frameworks, measurement, and possibilities for increasing various kinds of green space in cities.

NATURE IN THE CITY, REVISITED

RESOURCES

Cronon, W. (1991). *Nature's metropolis: Chicago and the Great West*. W. W. Norton.

Glaeser, E. L., & Kahn, M. E. (2010). The greenness of cities: Carbon dioxide emissions and urban development. *Journal of Urban Economics, 67*(3), 404–418.

What is the role of green space in cities? Are cities centers for dense populations with built physical infrastructure for compact living? Or are cities diverse, dynamic ecosystems where humans and nature interact on a daily basis? These views, while simplistic, are often debated in different forms, quite often with the representation of natural and built environments as a conflict, rather than in synergy. Perhaps a more apt line of inquiry focuses on the interconnectedness of the natural and built environment as a holistic socioecological system. As argued by William Cronon in *Nature's Metropolis:*

Chicago and the Great West, "The boundary between human and nonhuman, natural and unnatural is profoundly problematic" (Cronon, 1991, p. xix).

With over half of the world's population, cities not only are inseparable from nature but may present the best opportunity to achieve the transformative socioenvironmental change needed for global sustainability. It has been found that cities can be greener than suburban counterparts (Glaeser & Kahn, 2010) and have "a critical role to play in conserving biodiversity, protecting water resources, improving microclimate, sequestering carbon and even supply a portion of the fresh food consumed by urban dwellers" (Lovell & Taylor, 2013, pp. 1447–1448). Nevertheless, estimates indicate that cities, due to their status as home to the majority of the global population, are concentrated centers of global energy consumption and greenhouse gases emissions, while occupying less than 3 percent of global land (C40 Cities). Therein lies a conflict in that putting green space into cities, which have the least amount of available land but would directly affect the greatest number of people. This intensive use of a largely inelastic supply of land means that increasing the amount of urban green space will require creative use of distributed and highly functional spaces to provide myriad social and ecological benefits with equitable outcomes. The question remains whether nature and its benefits can flourish in cities, where human activity—a threat to nature—is concentrated, where our buildings use copious amounts of energy, where cars sit idling on hazy freeways and in downtown corridors. The answer requires an understanding of the benefits, trade-offs and mechanisms needed to provide equitable and high-functioning green space.

TOPIC 1: URBAN GREEN SPACE AND HUMAN HEALTH

RESOURCES

Jackson, L. E. (2003). The relationship of urban design to human health and condition. *Landscape and Urban Planning, 64*(4), 191–200.

Jennings, V., & Gaither, C. J. (2015). Approaching environmental health disparities and green spaces: an ecosystem services perspective. *International Journal of Environmental Research and Public Health, 12*(2), 1952–1968.

Tzoulas, K., Korpela, K., Venn, S., Yli-Pelkonen, V., Kaźmierczak, A., Niemala, J., & James, P. (2007). Promoting ecosystem and human health in urban areas using green infrastructure: A literature review. *Landscape and Urban Planning, 81*(3), 167–178.

Green space, from small on-site installations of greenery to large city or regional parks, has long been recognized as a way to create healthy urban environments. Green spaces provide healthy outdoor spaces for recreation and socialization, an ecological habitat for important species that live or migrate through urban areas, and important urban ecosystem services like clean air, heat mitigation, and flood control. Jackson (2003) details the benefits afforded by sustainable urban design at all city scales and the importance of exposure to green space for physical and mental health, as well as a social and cultural connection. Exposure to green space in cities has been shown to improve

GREEN SPACE

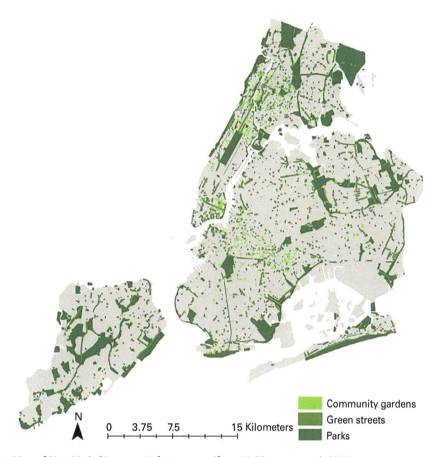

4.1 Map of New York City green infrastructure (from McPhearson et al. 2014).

human health and well-being in myriad ways, including through the facilitation of informal contact deemed important to healthy neighborhoods, as described by Jane Jacobs (1961).

But, the type and quality of green space varies, as does the size and distribution across cities (Cohen et al. 2010). Figure 4.1 from McPhearson et al. (2014) displays the spatial distribution of certain types of publicly provided green infrastructure throughout New York City. Since green infrastructure can take on many different forms, sizes, composition, and ownership, there lacks comprehensive data on citywide green space, but tools are rapidly emerging through remote sensing and mapping applications.

DISCUSSION QUESTIONS

1. The ecological and health benefits of green space in cities have been widely touted, as described above. So, why isn't there more green space in cities? What are the costs? And the trade-offs?
2. Should cities prioritize increasing green space in lower-income neighborhoods? Why or why not?

Box 4.1
Health Disparities and Access to Green Space

It has long been known that urban green spaces contribute to public health through direct and indirect pathways. Improved physical and mental health outcomes have been attributed to access to urban greenery and green spaces. Research has shown exposure to green space has been linked to an array of improved conditions, including reductions in surgical recovery times for those in hospital rooms with views of trees (Ulrich, 1984), decreased crime and violence (Bogar & Beyer, 2016), cognitive function and mental health in children and adults (Berman et al. 2008; Bratman et al. 2019; Dadvand et al. 2015), reduced asthma rates in children (Lovasi et al. 2008), participation in cultural activities (Dickinson & Hobbes, 2017), and overall health perception (Kardan et al. 2015; Maas et al. 2006).

Further, green space designed for walkability and recreation can provide an opportunity for improved physical conditions, including lower risk of cardiovascular disease and reduced obesity, as well as for social connections and community cohesion. The importance of having access to neighborhood green space became heightened during the COVID-19 pandemic, when people needing to get outdoors did not wander far from home. However, access to urban green space is not equal. And the type of green space matters. A study in cities in North Carolina found that park use became more homogeneous during COVID-19 and declined the most for those who were Black, Indigenous, or People of Color (BIPOC) or lower income in socially vulnerable communities, raising concerns about how existing health disparities could be exacerbated through inequitable access to parks and green space (Larson et al. 2021). Additionally, green space quality will differ by neighborhood due to the failure to engage community residents in the planning process, funding allocations and mechanisms, and investments in upkeep and maintenance.

Overall green space planning in cities is complex and multilayered. The decision rules, based on often misaligned goals related to access and ecological function, are often influenced by restrictions on land use or ownership, political representation, and funding. The well-known challenges to planners of achieving economic growth, environmental sustainability, and social justice, while perhaps not as polarizing or as simple as "jobs versus the environment," still remain as conflicting goals (Campbell, 1996, p. 297; Campbell, 2013). Advancing green space in cities, in particular, presents trade-offs including the connection between greenery and gentrification, as well as the loss in density or availability of land for other important needs as conservation measures increase. Creating green space in cities can involve both private and public actors, but it requires demonstrating nature's direct and indirect pathways to healthy urban living (Shanahan et al. 2015) while also paying careful attention to the measurability of its costs and benefits, as well as the distributional impacts of its outcomes.

Activity 4.1: Where Is the Green Space? (Homework)

Examine the green space in your city using city data, maps, observation, or other tools. Where is the green space? Is it evenly distributed? Is it near the city center or on the outskirts? If you have regional maps, do the areas outside of the city seem to have more green space? Given what you know about the city, or using census or health data, what can you say about the distribution of green space relative to population, neighborhood characteristics, demographics, or property values?

TOPIC 2: URBAN GREEN INFRASTRUCTURE

RESOURCES

Center for Neighborhood Technology. (2020). *Green values strategy guide: Linking green infrastructure benefits to community priorities.* https://cnt.org/sites/default/files/publications/Green%20Values%20Strategy%20Guide.pdf.

United States Environmental Protection Agency (U.S. EPA). (n.d.). What is green infrastructure? Retrieved February 2, 2021, from https://www.epa.gov/green-infrastructure/what-green-infrastructure.

Green space serves many purposes in cities but requires careful attention, protection, and planning to function as a source of sustainability and resilience. To orient nature as fundamental to the operation of cities, the term "green infrastructure" has been widely adopted to garner its importance in planning and investment decisions. Urban green infrastructure (UGI) includes protected natural areas like urban forests and wetlands, planned green spaces like parks and trails, and spaces in between and integrated into the built environment like green roofs and bioswales. Some definitions of urban green infrastructure also include built gray infrastructure that connects to natural systems, such as permeable pavement and stormwater detention ponds, or measures that facilitate better environmental outcomes and less reliance on gray infrastructure like downspout disconnections. While UGI is an encompassing set of natural, built, and hybrid systems, it has received the most attention and adoption for stormwater management in cities (Elliott et al. 2020). However, even while often built for specific purposes like stormwater, projects can be designed in ways that maximize the co-benefits afforded by UGI, including air quality control, regulation of climate, and habitat provision. This provision of benefits and co-benefits creates an alignment between UGI and the ecosystem services framework discussed in chapter 2, yet the location, design, and financing of UGI require considerations beyond ecosystem services.

Green infrastructure is defined under by the U.S. Environmental Protection Agency (U.S. EPA) Section 502 of the U.S. Clean Water Act as "the range of measures that use plant or soil systems, permeable pavement or other permeable surfaces or substrates, stormwater harvest and reuse, or landscaping to store, infiltrate, or evapotranspirate stormwater and reduce flows to sewer systems or to surface waters" (U.S. EPA, 2021). It includes both natural and humanmade approaches, usually in combination as a system that both employs and mimics nature, and is touted as a "cost-effective, resilient approach to managing wet weather impacts that provides many community benefits" (U.S. EPA, 2021). Cities have developed green infrastructure plans as complements to extensive gray infrastructure stormwater systems in places with combined sewer systems and vulnerability to urban flooding (City of Chicago, 2014; City of New York, 2019; City of Philadelphia, 2011). To implement such strategies, as well as ones at different scales, planning tools have become widely available to assess the suitability of green infrastructure types and to evaluate the associated benefits and costs.

The use of green infrastructure, both as a framework and a tool, goes beyond stormwater management and is broadly defined to include a range of urban applications. Research has demonstrated the benefits of different types of UGI, including large-scale land uses like parks (Sutton & Anderson, 2016), to distributed ones such as street trees and the urban forest (McPherson et al. 1997), green roofs (Yang et al. 2008), green alleys (Newell et al. 2013), and others. The functions and associated benefits of the many types of urban green infrastructure have been widely studied through various disciplinary lenses and at different scales (De Ridder et al. 2004; Hansen & Pauleit, 2014; Tzoulas et al. 2007). While modeling the function of different types of urban green infrastructure is largely done by ecologists and engineers, the benefits, costs, and trade-offs associated with land use have captured the attention of social scientists and urban planners.

Figure 4.2 from the Center for Neighborhood Technology shows the range of benefits associated with different types of urban green stormwater infrastructure. While all examples are related to the primary purpose of stormwater management, the co-benefits extend to priority areas for city and neighborhood planners.

Activity 4.2: Green Infrastructure Benefits and Costs for Decision-Making (Team Research and Class Discussion)

Figure 4.2 demonstrates the range of benefits from various types of urban stormwater green infrastructure. For each of these types, consider the benefits, costs, and feasibility of implementation and operation.

Benefits: For the noneconomic categories of benefits, consider how each benefit listed might be measured and monetized. Use scholarly articles, including references from this book, and calculators on the companion website to describe how dollar value estimates associated with the noneconomic benefits might be calculated. Provide some examples of the dollar value estimates for these benefits.

Costs: For the costs, beyond the cost of installation, what are the trade-offs or *opportunity costs*? Where might the money come to pay for such green infrastructure investments? You can recommend creative financing mechanisms (e.g., the savings associated with reduced energy costs could be used to plant trees for cooling purposes) or cite existing and innovative financing tools in practice. What are the trade-offs of such green infrastructure investments? Does it compromise a city's density goals or the compact city? How might it affect neighborhood diversity or violate transect rules? Would it lead to gentrification, and if so, is that necessarily bad?

Feasibility: Last, think about the feasibility of each type of UGI. How might it happen? Can you find any examples from cities that have been successful in implementing these types of UGI? Is community participation needed? Would cost be the main guiding principle? How would prioritization take place? Discuss your ideas.

GREEN SPACE

	GREEN STORMWATER INFRASTRUCTURE									
	Linear buffer park/trail	Stormwater park	Stormwater planter	Parkway bioswale	Rain garden	Street trees	Green roof	Permeable pavement	Permeable bike lane	District stormwater
HEALTH BENEFITS											
Improved outdoor air quality	•••	•	••	•	••	••	•			••
Improved indoor environmental quality	•	•••	••	•	•	•••	•	••	••	•
Reduced noise pollution	•••					•••		••	••	••
Reduced heat stress	••	•••	•	•	•	••	•			••
Improved community cohesion + mental health	•	•••	••	•	•	••	•			••
ECONOMIC BENEFITS											
Improved workforce development/job creation	•••	•••		•		•••	•	•	•	•••
Increased vacant land reactivation	•••	•••			•••					•••
Increased property values	••	•••	•	••	•	••	•••			•••
Increased sales revenue			••	••	••	••			••	••
Increased recreational revenue	•••	•••							•••	
CLIMATE ADAPTATION											
Reduced flooding	•••	•••	••	••	••	•••	••	•••	••	•••
Reduced urban heat island temperatures	••	•••	•	•	•	•••	•••	•	•	•••
Protected water quality (reduced runoff and combined sewer overflows)	•••	•••	••	••	••	•••	••	•••	••	•••
CLIMATE MITIGATION											
Reduced greenhouse gases	•••	•••	•••	•••	•••	•••	•••	•	•	•••
Reduced energy/fuel use	••	••				•••	•••			••
TRANSPORTATION BENEFITS											
Reduced on-street flooding	••	••	••	••	••	•••	•	•••	•••	•••
Improved safety	••		•	•					••	••
Increased opportunities for active transportation	•••	•••	••	•••	••				•••	••

••• High benefit   •• Medium benefit   • Low benefit

4.2 Community benefits of green infrastructure. *Source*: Table 1 from the *Green Values Strategy Guide: Linking Green Infrastructure Benefits to Community Priorities* (Center for Neighborhood Technology, 2020).

Activity 4.3: Green Infrastructure and Scale (Discussion)

Consider the following representation of green stormwater infrastructure in a city street from the Global Designing Cities Initiative. The green infrastructure system is designed to complement a traditional water drainage system put in place to reduce flooding at the street level by absorbing water and pollution before it enters the piped system (Global Designing Cities Initiative). The system is highly engineered to provide natural functions within the built environment. Comment on the pros and cons of such a system. Is this a vision of a sustainable future? Why or why not?

Box 4.2
Green Infrastructure and Planning Scales

Urban green infrastructure, green space, or built features that align with natural processes has emerged as a sustainable tool to complement aging gray solutions. UGI at site or local scales can contribute on the margin to reducing localized flooding or managing heat while generating important co-benefits for human and ecological health. Further, UGI as a planning tool may influence the design of cities toward ones that provide important co-benefits from additional ecosystem services, create awareness, and motivate changing ways of life. However, challenges to widespread adoption of UGI are related to definitions, densification, spatial scales, ownership, and equity.

Definitions: As described earlier, UGI can take on natural, engineered, or hybrid forms. While natural systems like urban forests or wetlands provide important co-benefits in the form of ecosystem services, they take up more urban space and have slower processes of environmental management. Engineered facilities can be managed toward specific environmental goals but lack the connections to community and associated cultural ecosystem service benefits. A more holistic and connected approach is necessary but requires different scales of planning and likely a new mind-set for city design.

Densification: The benefits of compact high-density cities have been detailed, but questions remain about the full relationship of density to environmental outcomes, including the challenges of creating new green space (Haaland & Konijnendijk van den Bosch, 2015). While innovations at the site and city scales have been implemented as a way to mitigate some of the unsustainable impacts of urban sprawl (Vargas-Hernandez & Zdunek-Wielgolaska, 2021) and contribute to well-being in already high-density cities, removal of green space in rapidly urbanizing places can occur due to new development. The process of densification can lead to fragmentation of local ecosystems and impede the holistic UGI planning needed for climate resilience, as well as exacerbate social inequalities. As the world continues to urbanize, the goal of a high-density compact city should be reconsidered with particular attention to holistic green space planning.

Spatial scales: Given the relative lack of space and fragmented landscapes in cities, UGI is generally small, distributed, and, at times, temporary. Green solutions can capture stormwater but, at the city scale, cannot replace gray storm and sewer infrastructure. Further, while trees provide a variety of important ecosystem services, dense tree canopy is difficult to achieve in cities, and trees are often removed for development or when disease hits. Further, the life span of carbon goes far beyond that of a tree, making it only a temporary storage facility. From a behavioral and design perspective, it is necessary to reconsider how cities are built to plan for urban resilience and sustainability. The question will remain if incremental change, such as that afforded by UGI, can meet the needs of a climate crisis at the global scale, or a much larger proposition is needed—one that, instead of considering nature in the city, considers the city within nature at the planetary scale. This type of transformation would require large-scale infrastructural investment, sociopolitical movements for change, and greater attention to equity as a guiding principle of sustainability (Goh, 2020).

Ownership: UGI can be prioritized based on need (e.g., stormwater management, heat mitigation), neighborhood or population, funding opportunities, or other measures. However, holistic planning will require involvement of all types of landowners. While it is easier to prioritize public land and buildings, much city land is privately owned and inseparable from an environmental perspective. While policies can support investment in UGI on private lands, the operation and maintenance is up to the private landowners, making long-term projects

> **Box 4.2**
> continued
>
> less viable. Planners should consider tools to make UGI the default or status quo, through zoning or policy changes, or revisiting city design to prioritize connected green infrastructure.
>
> *Equity*: There is a well-established correlation between urban green space access and predominantly white, affluent communities (Wolch et al. 2014). As a result, many cities have identified lower-income, minority neighborhoods for UGI, due to this lack of green space access, aging infrastructure, high risk to climate impacts, health disparities, social vulnerability, and, in some cases, more availability of vacant land. However, this comes with concerns about the relationship between urban greening and gentrification. While gentrification does not necessarily mean displacement, it is often a precursor to it and should be considered when making green infrastructure investments. This is not to say that certain neighborhoods should be deprived access to green space to avoid gentrification, but rather that planners should consider economic, environmental, and social outcomes of UGI planning (Wolch et al. 2014).

DISCUSSION QUESTIONS

1. Is green infrastructure a viable tool for urban planners? Is it the right framework to use to design sustainable cities? What are some of the pros and cons?
2. Consider the scale of urban green infrastructure. Can implementing UGI at a local or site-specific scale address a global problem like climate change?

> **Activity 4.4:** Green Infrastructure in Your Neighborhood (Homework)
>
> Find an example of green infrastructure in your neighborhood or city. You could walk around and look, then do some additional research. Or you could find a program online, then go visit a site. This can be at any scale, but it will be helpful to find a single site and then consider it as a citywide system such as a tree in an urban forest or a street in a network of green streets. Include a picture of the green infrastructure.
>
> Consider the following related to your type of green infrastructure:
>
> 1. Describe the green infrastructure resource. What is it? Where is it? Do people notice or interact with it? Take a picture if you can and include it in your writeup.
> 2. Who has provided this resource? Is it the city or other unit of government? A private landowner? Community group or nongovernmental organization? Or a combination? Comment on the scale of provision (single site, neighborhood, city wide, etc.).
> 3. Is there a particular program or policy that led to its provision? This could include tax credits, rebates, grants, mitigation fees, and development incentives. This tells us about the incentives needed for its provision.
> 4. Is this part of a larger network of green infrastructure? Does it connect to an ecological system in some way?
> 5. What ecosystem services does it provide? List the category and specific type of ecosystem services based on chapter 6, as well as the companion website resources. Who benefits from these ecosystem services?
> 6. How common is this type of green infrastructure? If very common, why is there so much of it? If not common, why isn't there more of it?

4.3 Green stormwater infrastructure. *Source:* Streetmix https://creativecommons.org/licenses/by-sa/4.0/.

TOPIC 3: PARKS AS URBAN GREEN SPACE

RESOURCES

Chiesura, A. (2004). The role of urban parks for the sustainable city. *Landscape and Urban Planning, 68*(1), 129–138.

Crompton, J. L. (2005). The impact of parks on property values: Empirical evidence from the past two decades in the United States. *Managing Leisure, 10*(4), 203–218.

Rosenzweig, R., & Blackmar, E. (1992). *The park and the people: A history of Central Park.* Cornell University Press.

Parks take on many different definitions and forms. Parks can be large, small, or even tiny patches in between buildings. Parks can be managed publicly by the federal government, states, counties, or municipalities; by private actors; or a hybrid model funded by private entities and operated by public ones. Parks can offer gathering spaces, sports facilities, children's play equipment, gardens, trees, native gardens, or other natural amenities, meaning the ecosystem services provided by a park can vary widely. The benefits of parks will vary as well, as will the accessibility, usability, and quality.

The value of green space in cities has long been studied, from early conjectures that people were willing to pay more for properties near parks to Frederick Olmstead's studies rationalizing the development of Central Park by estimating annual property tax revenues of nearly six times the annual costs of creating the park (Crompton, 2005). Economists developed more sophisticated methods to estimate the value of parks, including the hedonic property method (as mentioned in chapter 2), which focused on the incremental value attributed to park proximity, by controlling for all the other factors that influence real estate values.

Since property value approaches are premised on people living near the park, other approaches have been used to estimate the value of parks. For park visitors, the travel cost approach is used to determine how the time spent traveling and visiting a park reveals preferences and values for it. Because the value of park may also include those who do not live near it or visit it, approaches have also tried to capture the "existence" value of certain parks. Beyond strict economic approaches, integrated approaches from urban planning, landscape architecture, and engineering to assess the benefits of urban green space and infrastructure consider the dynamic relationship of green space to urban form (De Ridder et al. 2004), the role of participatory modeling and planning in supplying multifunctional urban green space (Lovell & Taylor, 2013), and spatial indicators to assess the distribution of green space in contrasting urban contexts (De La Barrera et al. 2016).

Yet, even with clear benefits afforded by park space in cities, there are concerns beyond those related to density and compact cities. Parks and urban green space are often mentioned as a source of gentrification and displacement. Historically, and still today, parks have been contested spaces, fraught with problems related to accessibility, crime, and sites of discrimination and racism. It has been well demonstrated that lower-income and minority neighborhoods lack green space, tree cover, and access to parks relative to wealthier, white neighborhoods (Jennings et al. 2012). As discussed in this chapter, this result is often the legacy of historical policies that have led to segregation, population loss, and systemic racism. The lack of green space also exposes residents in these neighborhoods to greater risks from flooding and heat exposure, leading to increased vulnerability. With careful attention, this may represent an opportunity. These neighborhoods tend to have greater amounts of vacant land, which, left as is, contributes to continued disinvestment. A vision for these areas reconstitutes the vacant land to advance urban greening through pocket parks, rain gardens, and urban farms.

The concern related to this vision is articulated to the use of property prices as a measure of the value of urban green space. Additionally, in practice, cities rely upon this premise as a tool for increasing revenue from property taxes. The challenge with such a reliance on increased property values is one of affordability. Often termed "gentrification," this becomes an environmental justice issue when increased or improved green space or park access displaces people in proximate neighborhoods. Historically, this has happened both directly as land was seized for parks or indirectly as long-time residents can no longer afford to live in neighborhoods after property values increase. One proposed solution is to make neighborhoods "just green enough" so that residents don't get displaced (Wolch et al. 2014). While this may reflect the only solution in a market-based housing system, the notion that people in lower-income neighborhoods must settle for less urban greening and therefore higher climate risk is inconsistent with the principle of environmental justice.

But what if the converse to gentrification is disinvestment? As coastal cities and climate sensitive areas become less attractive or higher-risk areas to live, people will move inland. This notion, centered on a fear-invoking discourse of climate migrants or climate refugees, leaves out the high vulnerability of those who cannot move or relocate. The displacement in these cases occurs then for those who are forced to stay in areas subject to high climate risk and disaster burdens.

As seen, the challenges to aligning environmental justice and equity with ecosystem services and green space goals emerge from the nature of the growth economy. A specific moral imperative would likely be necessary to counter or amend the strict cost-benefit rule so often used in public decision-making.

DISCUSSION QUESTIONS

1. How would you define a park? Does your definition require it to be a certain size, offer certain amenities or activities, be green, or be public?
2. The hedonic property method is based on the notion that people will pay more to live near a park, indirectly revealing their value for the park. Comment on if this approach fully captures the value of a park. What might be missing?
3. Do you think parks compromise density? Is this a problem, if so? Do parks contribute to gentrification? If this a problem, if so?

> **Activity 4.5:** ParkScore (Team Research)
>
> The Trust for Public Land publishes an annual index comparing park systems in cities called ParkScore. The index is calculated based on five categories: access, investment, amenities, acreage, and equity.
>
> 1. Visit the ParkScore website: https://www.tpl.org/parkscore/ and compare of the scores of two or three cities you are familiar with. Is this what you expected given what you know about these cities?
> 2. Describe how the index is calculated, including each of the five categories. Is anything missing, in your opinion? How would you rank the five categories in order of importance?
> 3. Why do you think ecosystem services are not included in the ParkScore? Do you think they should be?

At the same time, Central Park, in its origins, was a site of displacement. At a time when cities were expanding and industrializing, deteriorating environmental conditions led residents to seek refuge in open spaces. And so, a grand park resembling the public gardens of Europe was envisioned in 1800s by preeminent landscape architect Frederick Law Olmstead as a place for people to escape from the ails of the city, while not having to leave the city. However, there was no unused space in the tightly gridded streets of Manhattan, and the land was seized from residents, most notably leading to the destruction of Seneca Village, the city's first major settlement of African American property owners. The property value premiums, originally touted by Olmstead in justifying the construction of the park, also serve to make the surrounding neighborhoods unaffordable to many, if not most. Even today, Central Park remains a site of contested space, as is the case with many urban parks. In addition, some question whether the

> **Box 4.3**
> "The Lungs of the City"
>
> Central Park sits neatly in the gridded streets of Manhattan, comprising over 840 acres of municipal land in one of the most densely populated cities in the world (Levine, 2020). With nearly 40 million visitors per year, it is the most visited urban park in the United States. The ecosystem services provided by the park are on a grand scale, from its use as a large-scale recreation and gathering site, and a place for cultural amenities, including music performances, festivals, museums, and a zoo. In addition to its clear marketable value associated with tourism and proximate property values, the park provides myriad ecological benefits, including a diverse habitat for wildlife, mitigation of urban heat island effects, temperature and wind speed moderation, air quality improvement, carbon sequestration and storage, hydrological regulation and flood control, biodiversity protection, noise-level attenuation, and health benefits from increased physical activity, increased social interaction, and mental health (Sutton & Anderson, 2016).

(continued)

Box 4.3
continued

4.4 Central Park. *Source:* Photo Credit—Central Park Conservancy.

ecological properties of the Great Lawn, even with the innovative features, are truly sustainable. Nevertheless, Central Park is a widely used space by locals and visitors alike. Many cities have tried to emulate the park as a neighborhood gathering place and tourist attraction, as well as a valuable source of ecosystem services in an otherwise dense, urban, built environment.

Activity 4.6:   Central Park: The Greenest Place on Earth? (Debate)

In chapter 5, you will consider Times Square as the "greenest place on earth" due to its high-intensity urbanism and despite its lack of literal greenness. The proposition is that high density of the area maximizes social connection and keeps human settlement compact and contained, minimizing waste and sprawl. The counterproposition argued that human habitat required ample green space to provide ecosystem services, such as rainwater capture, habitat, and reduction of the urban heat island. Having green spaces in cities helps people connect with nature, providing a higher quality of life. Like Times Square, Central Park serves as a gathering and social space, but unlike Times Square, Central Park is nearly all green space with little to no residential density within the park boundaries. In this debate, students will consider the significant allocation of land for green space. Is the park in fact too green for a dense urban environment? Does it conflict with density goals and transect planning? Are large swaths of lawn space truly sustainable? Who are the beneficiaries of the ecosystem services provided by the park and who bear the direct and indirect costs?

The class will be divided into two teams to debate the following proposition: *"Central Park is the greenest place on earth."*

LITERATURE CITED

Berman, M. G., Jonides, J., & Kaplan, S. (2008). The cognitive benefits of interacting with nature. *Psychological Science, 19*(12), 1207–1212.

Bogar, S., & Beyer, K. M. (2016). Green space, violence, and crime: A systematic review. *Trauma, Violence, and Abuse, 17*(2), 160–171.

Bratman, G. N., Anderson, C. B., Berman, M. G., Cochran, B., de Vries, S., Flanders, J., Folke, C., Frumkin, H., Gross, J. J., Hartig, T., Kahn, P. H., Jr., Kuo, M., Lawler, J. J., Levin, P. S., Lindahl, T., Meyer-Lindenberg, A., Mitchell, R., Ouyang, Z., Roe, J., Scarlett, L., . . . Daily, G. C. (2019). Nature and mental health: An ecosystem service perspective. *Science Advances, 5*(7), eaax0903.

Campbell, S. D. (1996). Green cities, growing cities, just cities? Urban planning and the contradictions of sustainable development. *Journal of the American Planning Association, 62*(3), 296–312.

Campbell, S. D. (2013). Sustainable development and social justice: Conflicting urgencies and the search for common ground in urban and regional planning. *Michigan Journal of Sustainability, 1*(20181221), 75–91.

City of Chicago. (2014). Chicago green stormwater infrastructure strategy. https://www.chicago.gov/content/dam/city/progs/env/ChicagoGreenStormwaterInfrastructureStrategy.pdf.

City of New York, Department of Environmental Protection. (2019). NYC Green Infrastructure, 2019 Annual Report. https://www.1.nyc.gov/assets/dep/downloads/pdf/water/stormwater/green-infrastructure/gi-annual-report-2019.pdf.

City of Philadelphia, Philadelphia Water Department. (2011). Green city, clean waters. http://archive.phillywatersheds.org/doc/GCCW_AmendedJune2011_LOWRES-web.pdf.

Cohen, D. A., Marsh, T., Williamson, S., Derose, K. P., Martinez, H., Setodji, C., & Mckenzie, T. L. (2010). Parks and physical activity: Why are some parks used more than others? *Preventive Medicine, 50*(Suppl.), 10–13.

Crompton, J. L. (2005). The impact of parks on property values: Empirical evidence from the past two decades in the United States. *Managing Leisure*, *10*(4), 203–218.

Dadvand, P., Nieuwenhuijsen, M. J., Esnaola, M., Forns, J., Basagaña, X., Alvarez-Pedrerol, M., Rivas, I., López-Vicente, M., De Castro Pascual, M., Su, J., Jerrett, M., Querol, X., & Sunyer, J. (2015). Green spaces and cognitive development in primary schoolchildren. *Proceedings of the National Academy of Sciences of the United States of America*, *112*(26), 7937–7942.

De la Barrera, F., Reyes-Paecke, S., & Banzhaf, E. (2016). Indicators for green spaces in contrasting urban settings. *Ecological Indicators*, *62*, 212–219.

De Ridder, K., Adamec, V., Bañuelos, A., Bruse, M., Bürger, M., Damsgaard, O., Dufek, J., Hirsch, J., Lefebre, F., Pérez-Lacorzana, J. M., Thierry, A., & Weber, C. (2004). An integrated methodology to assess the benefits of urban green space. *Science of the Total Environment*, *334–335*, 489–497.

Dickinson, D. C., & Hobbs, R. J. (2017). Cultural ecosystem services: Characteristics, challenges and lessons for urban green space research, *Ecosystem Services*, *25*, 179–194.

Elliott, R. M., Motzny, A. E., Majd, S., Viteri Chavez, F. K., Laimer, D., Orlove, B. S., & Culligan, P. J. (2020). Identifying linkages between urban green infrastructure and ecosystem services using an expert opinion methodology. *Ambio*, *49*, 569–583.

Glaeser, E. L., & Kahn, M. E. (2010). The greenness of cities: Carbon dioxide emissions and urban development. *Journal of Urban Economics*, *67*(3), 404–418.

Goh, K. (2020). Planning the Green New Deal: Climate justice and the politics of sites and scales. *Journal of the American Planning Association*, *86*(2), 188–195.

Goulder, L. H., & Kennedy, D. (1997). Interpreting and estimating the value of ecosystem services. In P. Kareiva, H. Tallis, T. H. Ricketts, G. C. Daily, & S. Polasky (Eds.), *Natural capital: Theory and practice of mapping ecosystem services, nature's services. societal dependence on natural ecosystems* (pp. 15–33). Oxford University Press.

Haaland, C., & Konijnendijk van den Bosch, C. (2015). Challenges and strategies for urban green-space planning in cities undergoing densification: A review. *Urban Forestry & Urban Greening*, *14*(4), 760–771.

Hansen, R., Frantzeskaki, N., McPhearson, T., Rall, E., Kabisch, N. Kaczorowska, A., Kain, J., Artmann, M., & Paulet, S. (2015). The uptake of the ecosystem services concept in planning discourses of European and American cities. *Ecosystem Services*, *12*.

Hansen, R., & Pauleit, S. (2014). From multifunctionality to multiple ecosystem services? A conceptual framework for multifunctionality in green infrastructure planning for urban areas. *Ambio*, *43*(4), 515–529.

Jacobs, J. (1961). *The death and life of great American cities*. Vintage.

Jennings, V., Johnson Gaither, C., & Gragg, R. S. (2012). Promoting environmental justice through urban green space access: A synopsis. *Environmental Justice*, *5*(1), 1–7.

Kardan, O., Gozdyra, P., Misic, B., Moola, F., Palmer, L. J., Paus, T., & Berman, M. (2015). Neighborhood greenspace and health in a large urban center. *Scientific Reports*, *5*, 11610.

Larson, L. R., Zhang, Z. Oh, J., Beam, W. Ogletree, S. S., Bocarro, J. N., Lee K. J., Casper J., Stevenson K. T., Hipp, J. A., Mullenbach, L. E., Carusona, M., & Wells, M. (2021). Urban park use during the COVID-19 pandemic: Are socially vulnerable communities disproportionately impacted? *Frontiers in Sustainable Cities*, *3*, 1–15.

Levine, L. (2020, July 30). "The Lungs of the city": Frederick Law Olmsted, public health, and the creation of Central Park. *Gotham Center for New York City History*. https://www.gothamcenter.org/blog/the-lungs-of-the-city-frederick-law-olmsted-public-health-and-the-creation-of-central-park.

Lovasi, G. S., Quinn, J. W., Neckerman, K. M., Perzanowski, M. S., & Rundle, A. (2008). Children living in areas with more street trees have lower prevalence of asthma. *Journal of Epidemiology and Community Health, 62*(7), 647–649.

Lovell, S. T., & Taylor, J. R. (2013). Supplying urban ecosystem services through multifunctional green infrastructure in the United States. *Landscape Ecology, 28*(8), 1447–1463.

Maas, J., Verheij, R. A., Groenewegen, P. P., De Vries, S., & Spreeuwenberg, P. (2006). Green space, urbanity, and health: How strong is the relation? *Journal of Epidemiology & Community Health, 60*(7), 587–592.

McPherson, E. G., Nowak, D., Heisler, G, Grimmond, S., Souch, C., Grant, R., & Rowan, R. (1997). Quantifying urban forest structure, function, and value: The Chicago urban forest climate project. *Urban Ecosystems, 1*, 49–61.

McPhearson, T., Hamstead, Z. A., & Kremer, P. (2014). Urban ecosystem services for resilience planning and management in New York City. *Ambio, 43*(4), 502–515.

Newell, J. P., Seymour, M., Yee, T., Renteria, J., Longcore, T., Wolch, J. R., & Shishkovsky, A. (2013). Green alley programs: Planning for a sustainable urban infrastructure? *Cities, 31*, 144–155.

Shanahan, D. F., Lin, B. B., Bush, R., Gaston, K. J., Dean, J. H., Barber, E., & Fuller, R. A. (2015). Toward improved public health outcomes from urban nature. *American Journal of Public Health, 105*(3), 470–477.

Sutton, P. C., & Anderson, S. J. (2016). Holistic valuation of urban ecosystem services in New York City's Central Park. *Ecosystem Services, 19*, 87–91.

Tzoulas, K., Korpela, K., Venn, S., Yli-Pelkonen, V., Kaźmierczak, A., Niemela, J., & James, P. (2007). Promoting ecosystem and human health in urban areas using green infrastructure: A literature review. *Landscape and Urban Planning, 81*(3), 167–178.

Ulrich, R. S. (1984). View through a window may influence recovery from surgery. *Science, 224*(4647), 420–421.

United States Environmental Protection Agency (U.S. EPA). (n.d.). What is green infrastructure? https://www.epa.gov/green-infrastructure/what-green-infrastructure.

Vargas-Hernández, J. G., & Zdunek-Wielgołaska, J. (2021). Urban green infrastructure as a tool for controlling the resilience of urban sprawl. *Environment, Development and Sustainability, 23*, 1335–1354.

Wolch, J. R., Byrne, J., & Newell, J. P. (2014). Urban green space, public health, and environmental justice: The challenge of making cities 'just green enough'. *Landscape and Urban Planning, 125*, 234–244.

Yang, J., Yu, Q., & Gong, P. (2008). Quantifying air pollution removal by green roofs in Chicago. *Atmospheric Environment, 42*(31), 7266–7273.

5
DENSITY

Sustainable cities need to be compact in order to maximize access, minimize habitat loss, and promote social connection. This means that cities need to be "dense"—but how dense do they need to be? How can density be measured and evaluated, and what are the pros and cons of different levels and forms of density?

INTRODUCTION

RESOURCES

Jacobs, J. (1961). The need for concentration. In *The death and life of great American cities* (pp. 200–221). Vintage.

Sadik-Khan, J., & Solomonow, S. (2017). Density is destiny. In *Streetfight: Handbook for an urban revolution* (pp. 23–32). Penguin Books.

Keil, R. (2021). The density dilemma: There is always too much and too little of it. *Urban Geography, 41*(10), 1284–1293.

Sustainable cities need to be compact, which means they need to have density. Urbanists from Jane Jacobs to Janette Sadik-Khan have extolled the virtues of density (Jane Jacobs preferred the term "concentration") and the many positive rewards that dense urban living provides. For example, density can actually help reduce car traffic by making public transit feasible and more frequent. Crime can be reduced because density lends more "eyes on the street." If the concern is the blocking of light and air, there are ways to design and locate denser buildings in order to mitigate these problems. However, as will be discussed in this chapter, it is important to understand that density comes in many forms, and in order for density to have a positive assessment, the devil is in the details. Some forms of density are beneficial and delightful; other forms are not.

The benefits have been well researched. For example, research has shown that density correlates with protection of natural resources, lower CO_2 emissions, and social connection (Talen & Koschinsky, 2014). Compact, high-density cities are more resource efficient and consume less energy per capita, although a paradox is that there is often a higher concentration of environmental problems like pollution around densely developed areas. Per capita, however, the ecological footprint is lower. Studies have also shown that a simple doubling of standard suburban densities could significantly improve water quality (Jacob & Lopez, 2009), or that increasing density even slightly could yield millions of new homes (Romem, 2019), or that density makes walkability possible (Mooney et al. 2020). In fact, it is difficult to find contemporary literature that advocates for low-density, suburban, or rural development as a solution to climate change and housing problems.

Outside of the Western world, high-density is needed to accommodate massive surges in population. Developing or underdeveloped cities struggle to accommodate population growth, fueled by refugee crises, homelessness, and the spread of informal settlements such as slums and favelas. In the emerging economies of the world, addressing these problems simply cannot be a matter of low-rise development. China will grow cities for 300 million in the next twenty years, and the a priori assumption is that these cities will be very high density, mostly in the form of high-rises set in modernist superblocks.

Yet there are downsides, too. One is that density has been implicated as a factor perpetuating social injustices and inequities. One author made the distinction between *dominant density*—high-priced condominiums within or near downtown cores—and *forgotten density*: "favelas, shanty towns, factory dormitories, seniors' homes, tent cities, Indigenous reserves, prisons, mobile home parks, shelters and public housing." The problem with dominant density, which the author described as "a utopia of aspirational millennials and neat nuclear families with 1.5 children and a small hypoallergenic dog," is that it is a "privileged framework" that is oblivious to the health and social inequalities that forgotten density endures (Pitter, 2020).

The trick with density is figuring out what level and kind of density is needed to support sustainability goals. Here there is wide variation. Compare the examples shown in

---

Box 5.1
Density and the Pandemic

The global pandemic raised a question: is density a determinant in the transmission of COVID-19? According to researchers, the answer is no (Zhang et al. 2022). This is likely because higher-density cities are often wealthier and have the infrastructure and fiscal resources to combat public health emergencies. In addition, while there is some evidence that people with the means and ability to do so moved to the suburbs during the pandemic, the exodus has been exaggerated. What we have learned is that, even amid a global pandemic, all the technological enabling that allowed us to sprawl out, order online, and stay at home in front of our computers and televisions did not squash the human desire for density.

figures 5.1a and 5.1b. Figure 5.1a shows the high-rise Brickell neighborhood in Miami, Florida, which has a density of about 27,000 people per square mile. Figure 5.1b shows a section of the eleventh arrondissement of Paris, which has a density of about 110,000 people per square mile—almost four times the density of Miami's towers. Most buildings in the Paris image are less than eight stories (based on an analysis by Price, 2018).

Neighborhoods composed of single-family homes can actually be quite dense, depending on house and lot size. High-rise apartment buildings set in large swaths of open space can actually be low in density. In other words, large apartment complexes are a high-density building type only in certain contexts; when located in a park-like suburban context, they are a version of what Jane Jacobs ridiculed as "towers in a park" (and what modernist architects like Le Corbusier advocated). One truism about density is that to be valued and successful, density requires good urbanism—sufficient services and a well-designed public realm with buildings that contribute to a pedestrian-oriented context.

Activity 5.1: What Is Density and Where Should It Go? (Concept Map)

Some amount of human compactness is necessary for sustainability, but how much density should there be, when is it too much or too little, and where should it be located? For this concept map, teams will conceptualize the different parameters and considerations of population density. What are the pros and cons of different levels of density? The concept map should include parameters about (a) where places should be more dense, (b) where they should be less dense, and (c) what the implications are of this variation.

Teams should first attempt to define different types of density—start with high, medium, and low—and assign a specific density range to each definition. Then they should think about density in context and what level of amenities and services might be expected with each type of density.

Teams should evaluate how much of the city should be high density. Are there certain places where high density is absolute, such as near transit stations? What kind of form should this density take? Should cities be composed of high-rises as much as possible, or should all kinds of density be encouraged?

Teams should also consider who should live in the most dense places. Are high-rises appropriate for children? Should teenagers live in low-density contexts? Think about the kinds of housing units (from large single-family to studio apartments) associated with each kind of density.

TOPIC 1: DENSITY PREFERENCES

RESOURCES

Von Hoffman, A. (2021). Single-family zoning: Can history be reversed? *Housing Perspectives*. Joint Center for Housing Studies. https://www.jchs.harvard.edu/blog/single-family-zoning-can-history-be-reversed.

Wilkinson, W. (2019). The density divide: Urbanization, polarization, and populist backlash. https://www.niskanencenter.org/the-density-divide-urbanization-polarization-and-populist-backlash/.

58 CHAPTER 5

5.1a The Brickell neighborhood in Miami, Florida. *Source*: Google Earth.
5.1b The eleventh arrondissement of Paris. *Source*: Google Earth.

In the United States, the question of how much density cities need, and in what form, is highly controversial. Recent fights over the need for more density have pitted so-called "Nimby" groups (Not in My Backyard) against "Yimby" groups (Yes in My Backyard). There are groups formed around increasing density, like "Neighbors for More Neighbors" in Minneapolis, and there are groups working against density. One group, "Marin Against Density," or MAD, created an angry video railing against density ("The Story of How Marin Was Ruined"). Nimbys in the San Francisco Bay Area are well financed and politically active, as detailed in the article "Who Are the Bay Area's NIMBYs—And What Do They Want?" (Perigo, 2020).

An essential dilemma in the United States is that the American dream of a single-family house on its own lot is deeply ingrained in the American cultural consciousness. This can be a significant problem for sustainability if lots and house sizes are excessively large. For example, compare the lot and house size configurations of San Francisco, California, and Cave Creek, Arizona, shown in figures 5.2a and 5.2b. Both show a quarter square mile, but the density yield is dramatically different. San Francisco has 25-foot-wide lots yielding 480 single-family dwellings, while Cave Creek has 75-foot-wide lots yielding only 70 single-family dwellings. The higher density of San Francisco's lots has significant sustainability benefits. It allows services within walking distance of residents, sufficient ridership to sustain transit, and a level of compactness that sustains social connection, schools, street life, safety, and small businesses often dependent on pedestrians. And of course, through these effects, density has the ability to minimize car dependence, lower congestion and emissions, and promote active living.

The suburbs are densifying to some degree, and there is ingenuity in current efforts to retrofit them, for example, by putting housing on top of shopping malls (Williamson & Dunham-Jones, 2021). But absent these retrofitting efforts, the low-density of suburbia results in car dependency and the associated effects of pollution, congestion, and asphalt. The problem is that the large-format amenities of the suburbs—big boxes and "malls" of various kinds—only sustain their market by drawing from large swaths of the metropolis, which necessitates car travel. The density required to sustain services is simply not close at hand.

Land-use regulations are not helping the matter. In most zoning ordinances, density is controlled by specifying the amount of lot area required per dwelling unit. For single-family housing in cities, a common number is 5,000 square feet of lot area per dwelling. In the suburbs, it is not uncommon for this figure to be much higher—even a 2-or-more acre minimum. Sometimes this high acreage requirement is construed as an approach to land conservation, but unless housing is clustered in some way, the result is sprawl. Further, a high minimum lot area requirement is exclusionary, inequitable, and potentially racist. Growing recognition of this problem has motivated recent efforts to ban single-family zoning (although not single-family housing) outright, for example in Minneapolis, Connecticut, California, and Oregon (Von Hoffman, 2021).

5.2a San Francisco, California. *Source*: Google Earth.
5.2b Cave Creek, Arizona. *Source*: Google Earth.

> **Box 5.2**
> The "D" Word
>
> Dan Parolek, the author of *Missing Middle Housing*, suggests that conversations about density should avoid the "D" word. He writes, "Focus on increasing housing choice and attainability ... you will never convince any neighborhood that increasing density in and of itself is a good thing for their neighborhood. Consider terms such as Missing Middle Housing, house scale, housing diversity, and housing choices rather than density" (Parolek, 2020a).

Low-density housing pattern was not the norm in the early twentieth century. For example, consider the neighborhood design proposal of Clarence Perry, a sociologist in the early twentieth century who famously proposed a neighborhood scheme that was widely emulated (figure 5.3). His neighborhood was 160 acres (the acreage of a ½ mile square, within which Perry placed a circle with a ¼-mile radius), with a density of 10 units per acre and a population of 5,000—which, at the time, was the population needed to support an elementary school. A density of 10 units per acre well exceeds the typical density of an American suburb.

By the late 1950s, the superblock version of Perry's neighborhood unit arrived in the Soviet Union and China. It was very high density by American standards, double Perry's population size (5,000) on one-fifth of the acreage. The design was large-grid superblocks surrounded by arterials, with shopping at the periphery and quieter internal streets designed to keep out through-traffic. This kind of dense living is a far cry from the density most Americans seem to prefer, especially those inclined to suburban life.

Jane Jacobs (1961) suggested that negative perceptions about density might be more a matter of reacting to crowding rather than density per se. She argued that there was a need to make a distinction between crowding and concentration, arguing that crowding was an unhealthy condition that had to do with too many people per room. Concentration, on the other hand, was necessary for good urbanism. Jacobs advocated densities above 100 dwelling units per acre, which is very dense by American standards.

Activity 5.2: Housing Density from Tenements to Towers (Class Discussion)

The Skyscraper Museum in New York City created an exhibition contrasting the density level of twelve housing developments in the twentieth century. Students should read through the historical timeline before class, which highlights the density themes.

During class, students can be assigned to briefly overview four very different developments, all in New York:

Knickerbocker Village, 1934, 800 people per acre with 47 percent coverage
Williamsburg Houses, 1938, 210 people per acre with 36 percent coverage
Stuyvesant Town, 1947, 302 people per acre with 26 percent coverage
Manhattan House, 1951, 478 people per acre with 59 percent coverage

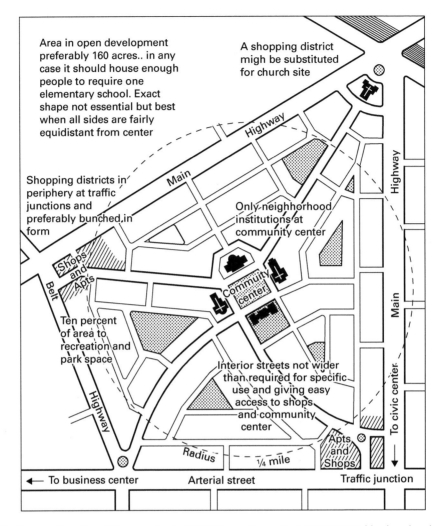

5.3 Clarence Perry's neighborhood unit, 1929. *Source*: Perry, C. A. (1929). *Neighborhood and community planning*. Regional Plan of New York and its Environs.

> These developments show the contrast between high density in the form of high-rises versus large-scale developments with a lot of surrounding open space. During discussion, the class can create a list of the comparative differences among these developments in terms of neighborhood context, affordability, connectivity and access, and demographics. Which density and form do students prefer? Which type seems to produce the best quality of life?

DISCUSSION QUESTIONS

1. Recent presidential elections in the United States have exposed a political divide that is predicted by density (Wilkinson, 2019). One associated critique is that lower-density areas tend to be socially uniform and adverse to diversity. Do you think this is the case?

2. Watch the seven-minute video "The Story of How Marin Was Ruined" by the group Marin Against Density (2014). Do you think valid points are being made?

TOPIC 2: LOCATING AND MEASURING DENSITY

RESOURCES

Cheng, V. (2009). Understanding density and high density. In E. Ng (Ed.), *Designing high-density cities: For social and environmental sustainability* (pp. 1–15). Routledge.

Lozano, E. E. (2013). Density in communities, or the most important factor in building urbanity. In M. Larice & E. Macdonald (Eds.), *The urban design reader* (pp. 399–414). Routledge.

LOCATING DENSITY

Sustainable cities should be dense, but where should density be encouraged? What parts of a city are the best places to try to increase density, without causing undo harm (such as higher housing costs)? One rule of thumb is that if services and transit are *not* evenly distributed, then it might make sense to calibrate density such that some neighborhoods have higher density and some have lower. On the other hand, if there is a more even distribution of services and transit, a more uniform density—such as that found in many Chinese cities—might be appropriate.

If there is an understanding of *where* an increase in density would be most suitable, density strategies can be tailored appropriately. This understanding is important because not every block can gracefully accommodate a higher-density building—different types of units are appropriate in different contexts. Some locations can increase density with a simple accessory building in the rear, while other locations can accommodate a full apartment block. It might make sense to have more density around public spaces like parks and civic institutions like libraries. Perhaps density should be higher near commercial areas in order to give the most people ready access to goods and services (if the commercial uses are compatible with housing).

Higher density close to public transit and around amenities and services makes sense because access to these elements makes high-density living worthwhile. In a sense, higher-density living, which is often accompanied by lower amounts of private space (especially outdoor private space), should be "rewarded" with greater access to public space and facilities. Of course, the reward needs to be worthwhile. There should be access to things worth having access to, like neighborhood-level services, transit stations, and valued civic spaces.

The idea that higher densities should be encouraged in places near public transit is based on several factors. The first has to do with parking. Sometimes residents object to allowing increased density because of a (real or perceived) increase in cars in the area. Tying the provision of more units to areas well served by transit can lessen this impact, as well as lessen the cost burden of car ownership. New units should be encouraged in areas close to transit—possibly even with parking restrictions instead of requirements. Another reason for increasing density near transit lines is that an increase in density supports public transport, resulting in higher ridership and, potentially, greater efficiencies.

> **Activity 5.3:** Density vs. Walkscore (Homework)
>
> Walkscore is an estimate of land-use diversity that measures the number of amenities within a given distance, weighted by category. If it's true that the benefit of living in a higher-density area is better access (especially walkable access) to amenities—the "reward" for high-density living—then we would expect to see a positive correlation between walkscore and density. Use walkscore (walkscore.com) and census data (census.gov) to compare the walkscores of different density levels for a city of your choosing. Do you find evidence that the relationship between walkscore and density holds? Make the case that this relationship does, or does not, seem to be in evidence. Include a table of your sampled walkscores and density levels.

MEASURING DENSITY

How can density be measured? There isn't a single right way. Most measures focus on the "hard" elements of density (population, housing units, land area), missing the "soft" side of density—human dimensions like behavior and perception, as well as context and quality (Boyko & Cooper, 2011). Some have argued that "raw density" is misleading and that a better measurement should consider the time and distance required for an individual "to encompass a zone suitable for the purposes of everyday life" (Nelson, 2016).

The simplest and by far the most common way of measuring density is to simply divide the population in a given area by land area. This is called "standard density." Not only is this exclusively a "hard" density concept, but it also can be misleading. It is the reason that Los Angeles might be calculated as being denser than New York City, which seems counterintuitive to anyone who has visited both cities. A better method is to calculate density by smaller areas and weight them based on population (see box 5.3). This gives an understanding of what kind of density most people are actually living in.

> **Box 5.3**
> Weighted Density vs. Standard Density
>
> Weighted density is different from standard density, as described in the blog post "Perceived Density." It is straightforward to make this calculation using 2020 census population and density (per square mile) by tract using data from the census.gov website. The basic method is to first select census tracts for a given city, then compute the density of each tract by dividing population by tract area and then assigning each tract a weight based on its percentage of the total population. In this way, sparsely populated tracts are discounted and densely populated tracts are weighted.

Another thing to watch out for is the difference between population density and housing unit density. These can be very different if the size of the housing unit varies. Depending on how many people live in a unit, a very different population density can result from a density measurement of dwelling units per acre (DUA).

An important distinction to be made in density measurement is between *gross* and *net* density. Gross density is population divided by the area of a whole spatial unit, with nothing factored out. Net density takes into account the land area that is not inhabited by housing units—like roads and public spaces. Hong Kong is a good example of how gross versus net density is consequential. As one author (Cheng, 2009) explained, "If the land area of the whole territory is taken into account, the overall population density in Hong Kong is about 6300 persons per square kilometre. However, only about 24 per cent of the total area in Hong Kong is built up. Therefore, if the geographical reference is confined to built-up land, then the population density will be about 25,900 individuals per square kilometre."

Activity 5.4: Density Differences (Team Research)

In class, teams will be assigned to places with different densities, such as Levittown (NY), Paris, Hong Kong, and the South Side of Chicago. One team could be assigned to investigate the density of a trailer park, using a SocialExplorer.com map called "Trailer Park City." Teams will research and estimate density levels.

Teams will decide on a uniform spatial unit (e.g., 1 square mile or 10 acres) so that results are more easily compared. Teams will share results with the class, comparing and contrasting density levels. Teams should first share images (without density calculations) so that the class can vote on which has the highest and lowest density. Were the results surprising? Teams should be prepared to discuss their measurement methods and discuss how they think density levels impact sustainability from an environmental, economic, and social point of view.

DISCUSSION QUESTIONS

1. The case is made that higher density should be close to amenities. What distance do you think is applicable? A five-minute walk? A ten-minute walk? How much would it depend on what the amenity is, like a grocery store versus a transit station?
2. Sometimes high density is located adjacent to recreational amenities—think of the wall of high-rises along the waterfront in Miami Beach. What should be done if this kind of density emerges in places vulnerable to climate change and rising sea levels?

TOPIC 3: DENSITY BY DESIGN

RESOURCES

Campoli, J. (2012). *Made for walking: Density and neighborhood form.* Lincoln Institute of Land Policy.

Campoli, J., & MacLean, A. S. (2007). *Visualizing density*. Lincoln Institute of Land Policy.

Parolek, D. G. (2020). *Missing middle housing: Thinking big and building small to respond to today's housing crisis*. Island Press.

Design is essential to density because it helps make density acceptable to people—even preferable. Good design can be used to show that density does not need to degrade urban life but can actually improve urban life. Good design can create better public spaces and provide support for the essential facilities and services people deem most important. It does this by helping to successfully mix different kinds of uses in a way that is cohesive and integrating rather than chaotic. Good design helps sustain a dense, walkable environment.

If density is associated with poorly designed urbanism, it will be fought against, which is an important instance of how design is essential to sustainability. For development in existing neighborhoods (infill), poorly designed higher-density housing might be seen as a threat to property values. This is why urban design is so critical—good design for infill housing can embed the infill within a larger context, making sure that it is possible to "envision each building, each development project, in relation to a positive ideal" (Brain, 2005, p. 32). In addition, research has shown that social problems sometimes associated with high-density living (crime, social isolation) can be improved through better design practices (Raman, 2010).

Julie Campoli and Alex MacLean made the case for the importance of good design as a critical component of density in their book *Visualizing Density* (2007); Campoli provided an update in her book *Made for Walking* (2012). The books argue that we can learn to love density by understanding that its negative effects—crowding and monotony—are the result of bad design, not density per se.

A recurrent idea about the best design for achieving density is that it should follow the form of low-rise but dense European cities. For example, the European architect Leon Krier (2015) makes the case that cities should be neither "land-scraping" nor "sky-scraping" but composed of compact two- to five-story buildings that are humanly scaled and do not require elevators. And in Yanarella and Levine's *Sustainable Cities Manifesto*, the authors argue that the virtues of social order, urban design, and political power are reflected in the typology of the medieval Italian hilltown: "high population density, a humanly scaled architecture, social heterogeneity, an urban-rural balance, primary political loyalties to the city as a whole, aesthetic richness and diversity of design and social commitment to durability and repair . . . [all] critical components of the modern sustainable city" (Yanarella & Levine, 2011, p. 24).

Relatedly, there is plenty of critique of density in the form of high-rise towers. One essay entitled "'Vertical Villages' May Be the Future of Urban Living. That's Scary" (Ehrenhalt, 2019) described high-rises as a kind of "dystopia." The good news, however, is that high density can be achieved without high-rises. As one writer summarized, "There are cheaper, more adaptable, and more economically inclusive development patterns

Box 5.4
The Stumpies

Density in the form of boxy apartment buildings (figure 5.4) has been much criticized. One explanation for the criticism is offered in the essay "Why America's New Apartment Buildings All Look the Same." Another blogger defended the building type, "In Praise of 'Boxy Buildings'"—five- to seven-story podium buildings—as being a good way to achieve density, mixed use, and affordability. But it is an approach to density that most neighborhoods resist—a type that has been pejoratively labeled a "Stumpie." One relevant question to resolve is whether very different housing types—single-family houses and high-rises—can be successfully mixed or whether this mixing brings a loss of neighborhood design quality. Is blocking "Stumpie" development motivated by a respect for the scale and character of an existing neighborhood, or is it about thwarting density and diversity? Watch "In Defense of the Gentrification Building" for an alternative view.

5.4 A "Stumpie" apartment block in Dallas, Texas. *Source*: Google Earth.

that achieve high population densities without having to jump straight into high-rise towers financed by big banks and built by huge development companies" (Price, 2018).

What was Jane Jacobs's preferred form of density? She thought that densities should be above 100 DUA and less than 200 DUA, arguing that "proper city dwelling densities are a matter of performance. . . . Densities are too low, or too high, when they frustrate city diversity instead of abetting it" (Jacobs, 1961, p. 208). The economist Ed Glaeser argued that Jacobs's targeted midrange density of 150 dwelling units per acre could be achieved with cities of six stories (Glaeser, 2010). However, one writer calculated that 150 DUA actually requires eight-story buildings at a minimum and that ten plus-story buildings are more realistic if you factor in parking needs (Orozco, 2012).

The design and form density takes are especially important when it comes to adding density in single-family neighborhoods—a key concern in the United States as some 67 percent of the U.S. housing stock is in the form of single-family houses. If designed well, multifamily infill housing can be compatible with single-family housing (figure 6.4 in chapter 6 shows the example of a fourplex that looks like a single- family house). Chapter 6 will discuss infill options in relation to adding diversity, but these same options are important for adding density, too. Increasing density can also be a matter of innovative types of multifamily housing—courtyards and closes (short, looped streets with housing around them) are examples. Early twentieth-century Garden City designers like Raymond Unwin were especially good at fitting in multifamily housing like attached row houses among single-family housing (figure 5.5).

"Missing middle housing" is a term used to describe housing that is multiunit but small in scale and therefore easily a part of single-family residential neighborhoods (Parolek, 2020b; examples of missing middle housing are provided on a companion website to Parolek's book *Missing Middle Housing* (www.missingmiddlehousing.com). It's called "missing" because this housing type (which Parolek defines as four- to eight-unit buildings) is in short supply—the United States is a country of single-family houses and, in larger cities, high-rises. Middle-level density housing has been declining. This is unfortunate because this middle scale of housing has important benefits: it can fit well within single-family neighborhoods, and it can provide a more affordable housing option.

Missing middle housing is all about increasing density by leveraging good design. With these well-designed housing types, higher-density housing can be added to a single-family neighborhood in a way that does not compromise the "character" of a neighborhood. In many parts of the United States, cities face enormous pressure to build more housing, especially in areas with transit access. As mentioned earlier, recent legislation in several states (e.g., California, Connecticut) has sought to force the issue, proposing to limit the right of municipalities to block midrise housing of moderate density in certain districts. Opposition by preservationists and affordable housing advocates has been intense, creating alliances and factions and pitting "Nimby" against "Yimby," environmentalist against developer, elected official against resident. Historic preservationists try to protect historic character; however, they are also viewed as elitists who are blocking change and the development of much-needed housing. Those against more development might simply be reacting to the relentless incursions of large, boxy four- and five-story apartments (see box 5.4) in neighborhoods of very different character.

DENSITY

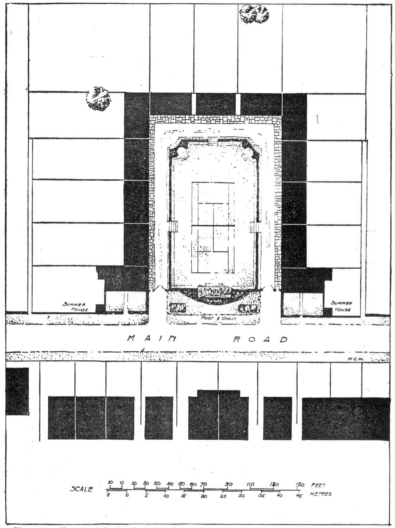

Illus. 277.—Hampstead Garden Suburb. Quadrangle of houses with carriage drive circling tennis-court, built for the Hampstead Tenants, Limited. See Illus. 292.

5.5 Raymond Unwin's courtyard housing, 1909. *Source:* Unwin, R. (1909). *Town planning in practice: An introduction to the art of designing cities and suburbs* (p. 354). T. Fisher Unwin.

The question is: can residents' fear about adding density be addressed through a more appropriate kind of design? Many residents don't want more density in their neighborhoods because it is thought to increase traffic, block light and air, increase crime, and depress property values. However, none of these objections are categorically true. It depends on the specific context—and design. If better design can be used to make a skeptical public more accepting of density, then much more attention needs to be paid to ensuring design quality.

DISCUSSION QUESTIONS

1. In your own experiences in the cities and towns you've lived in, can you identify examples of "missing middle" housing types? Do you think that the term is appropriate, in the sense that it seems to be "missing"?
2. Does the introduction of higher-density housing in a single-family neighborhood compromise the quality of a neighborhood?

Activity 5.5:   What Urban Form Is More Sustainable: Paris or Hong Kong? (Debate)

Calthorpe, P. (2015). Afterword: CNU expanded. In Congress for the New Urbanism and Emily Talen (Eds.), *Charter of the new urbanism* (pp. 253–257). McGraw-Hill.

Krier, L. (2015). Postscript: Completing the CNU charter. In Congress for the New Urbanism and Emily Talen (Eds.), *Charter of the new urbanism* (pp. 259–262). McGraw-Hill.

Density comes in many shapes and forms. But from a sustainability standpoint, is there a certain form of density that has the best hope of achieving sustainability goals like habitat protection, resource efficiency, and neighborhood diversity? Some urbanists have recently argued that density in the form of high-rises—for example, Hong Kong—is an unsustainable way to build cities. They argue that low-rise cities like central Paris are much more sustainable. Students can read the two essays above from the *Charter Book of the New Urbanism* to get a sense of the kinds of arguments being waged on both sides.

The class will be divided into two teams to debate the following proposition: *"Paris is more sustainable than Hong Kong."*

LITERATURE CITED

Brain, D. (2005). From good neighborhoods to sustainable cities: Social science and the social agenda of the new urbanism. *International Regional Science Review, 28,* 217–238.

Boyko, C. T., & Cooper, R. (2011). Clarifying and re-conceptualising density. *Progress in Planning, 76*(1), 1–61.

Ehrenhalt, A. (2019, March 19). 'Vertical villages' may be the future of urban living. That's scary. *Governing.* https://www.governing.com/archive/gov-vertical-villages.html.

Glaeser, E. L. (2010, May 4). Taller buildings, cheaper homes. *Economix Blog.* https://economix.blogs.nytimes.com/2010/05/04/taller-buildings-cheaper-homes/.

Jacob, J. S., & Lopez, R. (2009). Is denser greener? An evaluation of higher density development as an urban stormwater-quality best management practice. *Journal of the American Water Resources Association, 45*(3), 687–701.

Marin Against Density (MAD). (2014). The story of how Marin was ruined. https://youtu.be/qu8DBPx03T8.

Mooney, S. J., Hurvitz, P. M., Moudon, A. V., Zhou, C., Dalmat, R., & Saelens, B. E. (2020). Residential neighborhood features associated with objectively measured walking near home: Revisiting walkability using the Automatic Context Measurement Tool (ACMT). *Health & Place, 63*(2020), 102332.

Nelson, G. (2016). The deception of density. *CityLab*. http://www.citylab.com/tech/2016/10/the-deception-of-density/502646/.

Orozco, E. (2012, March 11). Proper scale: Squeezing Jane Jacobs into mid-rise urbanism. *Proper Scale*. http://properscale.blogspot.com/2012/03/squeezing-jane-jacobs-density-question.html.

Parolek, D. (2020a, February 25). Best practices for ending exclusive single-family zoning. *Public Square*. https://www.cnu.org/publicsquare/2020/02/25/best-practices-ending-exclusive-single-family-zoning.

Parolek, D. G. (2020b). *Missing middle housing: Thinking big and building small to respond to today's housing crisis*. Island Press.

Perigo, S. (2020, February 20). Who are the Bay Area's NIMBYs—and what do they want? *Curbed San Francisco*. https://sf.curbed.com/2020/2/20/21122662/san-francisco-bay-area-nimbys-history-nimby-development.

Pitter, J. (2020). Urban density: Confronting the distance between desire and disparity. *Azure Magazine*. https://www.azuremagazine.com/article/urban-density-confronting-the-distance-between-desire-and-disparity/.

Price, A. (2018). Surprising approaches to achieving density. *Strong Towns Journal*. https://www.strongtowns.org/journal/2018/1/3/comparing-approaches-to-achieving-density.

Raman, S. (2010). Designing a liveable compact city: Physical forms of city and social life in urban neighbourhoods. *Built Environment (1978–), 36*(1), 63–80. http://www.jstor.org/stable/23289984.

Romem, I. (2019). A modest proposal: How even minimal densification could yield millions of new homes. *Zillow Research*. https://www.zillow.com/research/modest-densification-new-homes-25881/.

Talen, E., & Koschinsky, J. (2014). Compact, walkable, diverse neighborhoods: Assessing effects on residents. *Housing Policy Debate, 24*(4), 717–750.

Williamson, J., & Dunham-Jones, E. (2021). *Case studies in retrofitting suburbia: Urban design strategies for urgent challenges*. Wiley.

Yanarella, E., & Levine, R. (2011). The Sustainable Cities Manifesto. In *The city as fulcrum of global sustainability* (pp. 23–40). Anthem Press.

Zhang, X., Sun, Z., Ashcroft, T., Dozier, M., Ostrishko, K., Krishan, P., McSwiggan, E., Keller, M., & Douglas, M. (2022). Compact cities and the Covid-19 pandemic: Systematic review of the associations between transmission of Covid-19 or other respiratory viruses and population density or other features of neighbourhood design. *Health Place. Jul;76:102827*.

6

NEIGHBORHOOD DIVERSITY

A fundamental goal of urban sustainability is that neighborhoods should be socially and economically diverse—mixed in income, mixed in use, and actively supportive of places that commingle people of different races, ethnicities, genders, ages, abilities, occupations, and households. How realistic is such a goal? What level of mix currently exists, and how much is needed? In this chapter, students will evaluate diversity goals, explore methods of diversity measurement, and think about the value and practicality of supporting diversity through policy and zoning reform.

INTRODUCTION

RESOURCES

Jacobs, J. (1961). The kind of problem a city is. In *The death and life of great American cities* (pp. 428–448). Vintage.

Florida, R. (2017). The diversity–segregation conundrum. *American Journal of Community Psychology, 59*(3–4), 272–275.

Sampson, R. J. (2019). Neighbourhood effects and beyond: Explaining the paradoxes of inequality in the changing American metropolis. *Urban Studies, 56*(1), 3–32.

Diversity—the mixing of people and uses—is a fundamental component of urban sustainability. In fact, the city is revered precisely because it is the locus of difference and diversity, constituting a positive force in a global society and a mode of existence that enhances human experience. The writings of Fischer (1975), Lefebvre (1991), and Harvey (2000), among others, have explored the ways in which diversity sustains human life and how the lack of diversity—separation and segregation—undermines sustainability. This mirrors a fundamental principle of human ecology: diversity is what builds resilience in both human societies and natural ecosystems (Capra, 1996).

Historically, urban neighborhoods used to be much more diverse, at least in terms of income and occupation (Talen, 2018). For example, historians used taxes and rents to show a high level of wealth and occupational variation at the neighborhood level in fifteenth-century Florence (Eckstein, 2006). Another study of residential segregation in the sixteenth to nineteenth centuries in Europe was able to show a complex interweaving of social groups based on data drawn from tax records, rents, and known locations of "recipients of poor relief" (Lesger & Van Leeuwen, 2012). The nineteenth-century social reformer Charles Booth created maps showing the streets of London's upper classes living adjacent to streets inhabited by people "of chronic want," attesting to a high level of social mixing still evident in late nineteenth-century London (figure 6.1). This does not mean that people of varying social status had an equalized view of each other, but it does mean that they shared the same public realm and had the same level of access to what the city had to offer.

Unfortunately, many cities, over the course of the past century, have become increasingly segregated, both in terms of separating land uses and in terms of separating people. For American cities in particular, the enduring reality is that they are highly segregated, racially, ethnically, and economically. While the suburbs generally have become less white and less affluent in recent years, these population shifts have not reduced neighborhood-level social segregation. In fact, the racial and economic gaps between city and suburb, or between one suburb and another, or between one neighborhood and another, have widened in the past half-century (Owens, 2015; Reardon et al. 2015). In the United States, the spatial clustering of "likeminded America" intensifies each time a family moves, creating what Bill Bishop (2009) called the "Big Sort."

These trends run counter to sustainability goals. When people and land uses are segregated (i.e., not mixed), the result is most often inequity: some people will have excellent access to goods and services, and others will not. A sustainable neighborhood is a neighborhood where residents have equal access to essential facilities and services, no matter what neighborhood they live in (Beatley & Manning, 1997). Neighborhoods that, in addition to being mixed in goods and services, are also socially diverse provide access to a wider range of socioeconomic contexts. Diversity in all forms, economic and social, is especially important for locally oriented populations—residents who rely on modes of transport other than the automobile.

Beyond the question of access, the inability to deal with social difference at a localized level—at the scale of a neighborhood—is seen as the root cause of intractable urban issues from sprawl and inner-city disinvestment to failing schools and environmental degradation. Lack of land-use diversity at the neighborhood scale translates to car dependency, which is the hallmark of an unsustainable city. A mixture of land uses has been shown empirically to encourage non-automobile-based modes of travel such as walking and bicycling, which in turn are seen as having a positive impact on public health (Frank et al. 2006). It is ironic that the environmental crisis of cities a

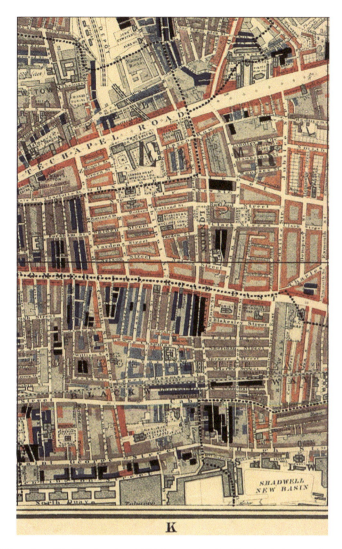

6.1 Map of Whitechapel in East London, 1889 *Source*: Booth, Charles. (1889). *Labour and Life of the People. Volume 1: East London.* London: Macmillan. Street colors represent the economic class of residents: Yellow ("Upper-middle and Upper classes, Wealthy"), red ("Lower middle class - Well-to-do middle class"), pink ("Fairly comfortable good ordinary earnings"), blue ("Intermittent or casual earnings"), and black ("lowest class . . . occasional labourers, street sellers, loafers, criminals and semi-criminals"). *https://commons.wikimedia.org/wiki/File:Commercial_Road,_Whitechapel,_Charles_Booth,_Map_Life_London_nek56.jpg https://upload.wikimedia.org/wikipedia/commons/e/eb/Commercial_Road%2C_Whitechapel%2C_Charles_Booth%2C_Map_Life_London_nek56.jpg* Public domain, via Wikimedia Commons century ago was remedied by separating uses (residential from industrial, for example), a strategy that later proved to generate its own set of adverse environmental effects.

Lack of income and racial diversity translates to neighborhoods that experience disinvestment, lack of services, lower environmental quality, and an overall reduction in support for public investment in neighborhoods that are less well off (Sharkey, 2013; Talen & Lee, 2018). Social segregation plays a clear role in exacerbating environmental harms and injustices (Harris et al. 2020), and lack of racial and economic integration promotes concentrations of wealth that hoard resources and concentrations of poverty that aggregate disadvantage (Galster, 2019). It follows, then, that sustainability theorists envision the sustainable city as one where separation by income or race is minimized.

Racism and classism have certainly played a part in creating segregated cities, aided by government actions. During the past century in particular, one of the most noticeable failures of housing policy at all levels has been the fostering of human separation into different types of living environments. In his book *The Color of Law: A Forgotten History of How Our Government Segregated America*, Richard Rothstein (2018) exposed the many ways that land-use regulation segregated African Americans and kept them out of certain neighborhoods. Other forces have been at play—some voluntary, some compulsory, some opportunistic—but the end result is the geographic separation of population subgroups throughout metropolitan regions.

But now, the idea of *diversity*—the antidote of separation and segregation—has become a key goal in urban sustainability. Urbanists now often argue that neighborhoods should be socially and economically diverse—mixed in income level, racial makeup, and land use. New Urbanists, smart growth advocates, creative class adherents, sustainability theorists—all have espoused the fundamental goal that a diversity of people and functions should be spatially mixed.

Jane Jacobs (1961) advanced this principle, arguing persuasively that diversity is an essential ingredient of healthy cities. What counted for Jacobs was the "everyday, ordinary performance in mixing people," forming complex "pools of use" that would be capable of producing something greater than the sum of their parts (pp. 164–165). Earlier, Lewis Mumford (1938) had stressed the importance of social and economic mix: a "many-sided urban environment" was one with more possibilities for "the higher forms of human achievement" (p. 486). He also wrote that the physical design of cities was supposed to foster this mix wherever possible to achieve the mature city: "A plan that does not further a daily intermixture of people, classes, activities, works against the best interests of maturity" (Mumford, 1968, p. 39). The solution is to foster a "close-grained" diversity of uses that provides "constant mutual support," whereby the focus is on, as Jacobs (1961) put it, "the science and art of catalyzing and nourishing these close-grained working relationships" (p. 14).

There are some interesting parallels between diversity in cities and diversity in the biological world. Since ecological health is often a matter of sustaining plant and

Box 6.1

Mapping Inequality

"Redlining" refers to the maps that were created by agents of the federal government's Home Owners' Loan Corporation (HOLC) between 1935 and 1940 (figure 6.2). The HOLC used information from local real estate appraisers, lenders, and developers to assign grades that they thought reflected a neighborhood's "mortgage security." Those receiving the lowest grade of "D" were colored red and were considered "hazardous." One factor of the appraisal was whether a neighborhood was "heterogeneous," in which case it would receive a lower grade.

An excellent way to explore the legacy of redlining in American cities is through the interactive map *Mapping Inequality: Redlining in New Deal America*, part of the "American Panorama" website, a digital atlas of American history (Nelson et al. n.d.). The project is the result of a collaborative effort among a group of historians and digital humanities scholars at four universities.

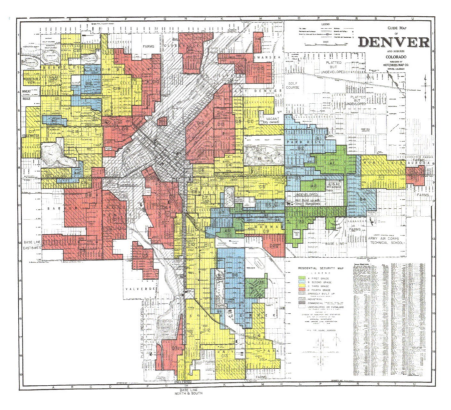

6.2 An HOLC "redlining" map of Denver, Colorado. HOLC maps are in the public domain. Creative Commons Attribution-NonCommercial-ShareAlike 4.0 International License. *Source*: Nelson, R. K., Winling, L., Marciano, R., Connolly, N., et al. (n.d.). Mapping inequality. In R. K. Nelson & E. L. Ayers (Eds.), *American panorama*. Retrieved May 31, 2022, from https://s3.amazonaws.com/holc/tiles/CO/Denver/1938/holc-scan.jpg.

animal diversity, concepts about maintaining habitat and species diversity help us understand the human habitat as a diversity-sustaining place. This does not require drawing a direct link between biological and social diversity, since humans behave in ways that are much different from plants and animals. But just as landscape ecology explores the connections between landscape design and ecological sustainability (Dramstad et al. 1996), we can explore the connections between the urban environment and its ability to sustain diversity.

DISCUSSION QUESTIONS

1. Do you think diverse neighborhoods are fundamental to urban sustainability? Explain why you believe there is or is not a fundamental relationship.
2. What do you see as the limits of neighborhood diversity? Are there certain dimensions on which homogeneity seems desirable, such as family type and age? What about race and ethnicity?

Activity 6.1: What Is Neighborhood Diversity? (Concept Map)

How should neighborhood diversity be defined? Two things need to be considered: the spatial area to be used to measure diversity and the variables to be used to characterize diversity. The first question involves conceptualizing what defines a neighborhood—a few blocks? a few acres? a census tract?—and how variation of the size of neighborhood (larger or smaller) changes the calculation and understanding of diversity. The second question is about the social or other variables to be used to define diversity. Is diversity only about people, or is it also about matters of place, such as building type, architectural style, or the variety of retail options? Is diversity of age or family type, rather than income or race, a sufficient way to define diversity?

Working in teams, create a concept map of these diversity parameters and how they interrelate. Teams should start by discussing the wide array of diversity measures and constructs involved. What are the different purposes of diversity, and why is it important to urban sustainability? What are the different ways it can be defined, and what are the connections among these variations?

TOPIC 1: SOCIAL MIXING

RESOURCES

Jacobs, J. (1961). The generators of diversity. In *The death and life of great American cities* (pp. 143–151. Vintage.

Darrah-Okike, J., Harvey, H., & Fong, K. (2020). 'Because the world consists of everybody': Understanding parents' preferences for neighborhood diversity. *City & Community, 19*(2), 374–397.

Although a basic tenet of urban sustainability is that diversity of people and of land uses within a neighborhood is essential, figuring out what that means in practical terms is

NEIGHBORHOOD DIVERSITY 79

> **Box 6.2**
> Rich and Poor, Side by Side
>
> Take a look at the dramatic differences between where rich and poor people live in aerial photos from South Africa, Mexico, and India on the website https://unequalscenes.com/. The images show starkly different living experiences in very close proximity—rich and poor side by side. Does this close proximity constitute a form of "diversity"?

no small task. What level and type of diversity should be targeted, and at what scale? And if society "naturally" gravitates toward less diversity over time, aided especially by market forces and housing preferences, should governments try to stop this tendency?

Not surprisingly, neighborhood diversity—especially if targeted as a proactive policy goal—has its critics. Some argue that social mixing does more harm than good by creating a situation in which more affluent neighbors stimulate resentment among less-affluent neighbors. There are claims that social mix policies harbor a "hidden social cleansing agenda," where words like "renaissance," and "sustainability" stand in for gentrification (Lees, 2008). Where an increased social mix is the result of white residents moving into African American neighborhoods, longtime (and often less-affluent) residents might feel alienated or resentful. In his book *There Goes the 'Hood* (2006), Lance Freeman showed how, after a history of disenfranchisement, redlining, and discrimination, an appreciation of neighborhood improvements via "social mix" was understandably met with a certain degree of cynicism toward social mix via gentrification.

Efforts to create mixed-use developments that combine economic and social functions also have positive and negative effects. On the positive side, mixing improves access because it minimizes the distances between people and goods. But such mixing might be resisted as an inauthentic attempt to contrive a false type of urbanity, an attempt to commodify a nostalgic notion of what pedestrian-based cities used to be like. As services change in response to a new demographic, long-timers might miss the homegrown services their neighborhoods used to have.

Perhaps the most significant debate about neighborhood diversity is whether governments should be in the business of proactively engendering social diversity. Should neighborhood diversity be supported as a matter of policy? Proponents argue that just because diversity is challenging—requiring strategies to counteract homogenizing forces as well as difficult choices about the level and scale of diversity to be targeted—this does not mean it should be abandoned as a matter of proactive policy. There is too much to be gained: the creation of socially diverse neighborhoods is one of the most important conditions of a vital and well-functioning (i.e., sustainable) human settlement. Because of the many benefits, urban diversity should be encouraged wherever and however possible. Governments need to push back against homogenizing social and market tendencies and try to enable, wherever feasible, different kinds of uses and different kinds of users who can intermingle in regular contact throughout the day. And remember, proponents say, human diversity is not only the basis of social equity but also an economic asset.

A counterresponse is that separation is human nature: people like to be with their "own kind." This means that the construction of socially mixed neighborhoods is contrived and artificial. Constructing affordable housing in high-income neighborhoods—or high-income housing in poor neighborhoods—entails a significant misunderstanding about how people actually relate to each other. Low-income residents do not benefit from high-income neighbors as models of behavior, and in any case, higher-income residents tend not to provide social support via organizational involvement (Chaskin & Joseph, 2015).

In fact, economists argue that social segregation occurs for a variety of well-documented—and justified—reasons. These include the Tiebout hypothesis (social homogeneity is based on levels of taxes and services), Schelling's hypothesis (households desiring at least some similar neighbors will result in widespread segregation), the effect of topography, the utility interdependence of neighbors, and the external economies generated by socially homogeneous areas (access to goods from a scattered population will result in high transport costs). Sociologists have theorized that social homogeneity strengthens social support networks, helps protect against discrimination, and helps to preserve cultural heritage (Suttles, 1972). Vale (2013) maintains that income mixing might sound good politically, but it is socially unnecessary.

DISCUSSION QUESTIONS

1. Do you believe it's the responsibility of government to change the rules of city building such that diversity is supported? If so, how far should society go in trying to support them?
2. Do you worry that proactive support on the part of government seems like social engineering?

TOPIC 2: MEASURING DIVERSITY

RESOURCES

Talen, E., & Lee, S. (2018). Mix. In *Design for social diversity* (pp. 107–129). Routledge.

Wright, R., Ellis, M., Holloway, S., & Golriz, M. (2021). Mixed measures: Different definitions of racially diverse neighborhoods compared. *Urban Geography*, *42*(8), 1147–1169.

Depending on how diversity is defined, every city has at least some neighborhoods that are diverse. Mixing housing types and sizes within neighborhoods is one contributing factor (Talen & Anselin, 2022). In addition, diverse neighborhoods are often physically older than nondiverse neighborhoods in terms of building stock. This is because older places are more likely to have experienced a housing filtering process, whereby some proportion of its older housing stock has become more affordable. Zoning also plays a role by allowing mixed housing types and mixed uses, as well as by eliminating rules that homogenize development (requirements for minimum lot size, maximum density, minimum setbacks, and other barriers to diverse development types).

A diverse neighborhood is a place that harbors a full range of human complexity, but what that consists of in a practical sense is open to interpretation. People consider the mixing of residents by race/ethnicity and by income level or wealth to be the most essential criteria, although mixing by age and family type, ability/disability, or religion might also be important aspects. A diverse neighborhood may have teenagers and elderly, married couples and singles, empty nesters and large families, waiters and merchandizers, as well as professionals, hedge funders, and people on fixed incomes. A diversity of people requires a diversity of uses that can support and sustain the mix.

Measuring diversity or mix involves understanding how different groups or types of land uses are clustered within a defined area. Thus, it is necessary to first define *what* a neighborhood is (i.e., its spatial extent). Some cities have official designations, although many others do not (Chicago, for example, the "city of neighborhoods," has no official neighborhood designations). Sometimes neighborhood definition is based on historical tradition. A painfully uncomplicated definition of neighborhood was offered in a 1957 editorial published in the *New York Post*: "A neighborhood is where, when you go out of it, you get beat up" (Kempton, 1963). Social scientists often define neighborhoods based on census tracts.

Neighborhood definition might be based on physical factors such as changes in street pattern, block size, or platting pattern. Planners have defined neighborhoods in terms of the walking distance between where people live and the goods and services they need on a daily basis, perhaps an area with a ¼- or ½-mile radius. Back in the early part of the twentieth century, neighborhoods were defined as a ¼-mile circle around a school. Or they might be oriented around commercial nodes, transit stops, or intersections conceived as neighborhood centers. These nodes might be used to construct a "pedestrian shed"—the five- or ten-minute (¼- or ½-mile) walk around the central place. Finally, neighborhoods might be delineated based on strong edges that are either humanmade or natural (e.g., thoroughfares, rail lines, parks, waterways).

In terms of measuring diversity, deciding on an appropriate spatial unit—a neighborhood—is the hard part. Once defined, then the range of groups and land uses within a neighborhood can simply be counted up and compared. This range of diverse elements might be comparatively low or high: homogeneous and lacking in diversity or heterogeneous and diverse.

DISCUSSION QUESTIONS

1. Review the official or quasi-official neighborhood maps for your city. What do you think are the determinants of their boundaries: historical, political, socioeconomic?
2. To what degree are the neighborhoods in your city defined by social similarity?

Activity 6.2: Is Your Neighborhood Diverse? (Homework)

Select a neighborhood to investigate for its diversity—perhaps the area where you grew up or around your current place of residence. The neighborhood should be large enough to cover at least one census tract. Do you think it's diverse, and in what sense—by race, ethnicity, income, family type, age? Use census data to analyze the diversity of your neighborhood. A good place to retrieve census information is the website socialexplorer.com. Select variables that you believe can be used to define diversity—race, ethnicity, and median income would be the most common variables to use. Download tract data as an Excel or csv file and use the Simpson Diversity Index (see box 6.3) to calculate tract diversity. In class, students can present their findings and compare the diversity of their neighborhoods. Whose neighborhood is the most diverse?

TOPIC 3: MIXED-INCOME HOUSING

RESOURCES

Chaskin, R. J., & Joseph, M. L. (2015). Concentrated poverty, public housing reform, and the promise of integration. In *Integrating the inner city: The promise and perils of mixed-income public housing transformation* (pp. 3–22). University Chicago Press.

Freidrichs, C. (Director). (2011). *The Pruitt-Igoe Myth* [Film; DVD release]. https://weta.org/watch/shows/america-reframed/pruitt-igoe-myth.

Talen, E. (2021). Ten urban design strategies for fostering equity and inclusion in mixed-income neighborhoods. https://case.edu/socialwork/nimc/sites/case.edu.nimc/files/2020-07/Talen.WWV_.Ten%20Urban%20Design%20Strategies.2020.pdf.

Box 6.3
Simpson Diversity Index

There are many diversity indices. We'll use the Simpson Diversity Index, which is straightforward to calculate and interpret. The following formula can be easily entered into a spreadsheet.
Enter the following formula in a spreadsheet cell:

Diversity = (A3 * (A3 – 1)) / ((B3 * (B3 – 1)) + (C3 * (C3 – 1)) + (D3 * (D3 – 1)) + (E3 * (E3 – 1)) + (F3 * (F3 – 1)))

where

- A3 = total population in all categories combined
- B3 = total in category 1 (e.g., non-Hispanic white)
- C3 = total in category 2 (e.g., non-Hispanic Black)
- D3 = total in category 3 (e.g., Hispanic)
- E3 = total in category 4 (e.g., Asian)
- F3 = total in category 5 (e.g., all others)

A high diverse tract will have a number closer to the total number of categories used. For example, if there are four categories and the diversity index is "4," then the population is evenly distributed into the four categories. If the diversity index is "1," then the population is all in one category and therefore not diverse.

6.3 North Town Village in Chicago, Illinois, a HOPE VI mixed-income development. *Source*: Google Earth.

The U.S. federal government launched a program called "HOPE VI" in the 1990s as a new approach to public housing: it was to be mixed in income. In these developments, there is some public housing, some lightly subsidized affordable housing, and some fully market-rate housing, built using a range of unit types. North Town Village in Chicago, for example, constructed on the site of a former high-rise public housing project known as Cabrini-Green, has 50 percent market-rate units, 20 percent affordable units, and 30 percent public housing units reserved for former residents of Cabrini-Green (figure 6.3).

Although mixed-income housing is still a preferred public policy, HOPE VI and other subsequent mixed-income housing developments have been heavily criticized. One issue is that mixed-income developments are just that—mixed in income but usually racially and ethnically homogeneous. And the attempt to mix incomes feels, to some critics, "contrived." Spotlighting the social dynamics of income mixing in reconstructed ("transformed") public housing projects in Chicago, Chaskin and Joseph (2015) exposed the many perils of what would seem on the surface to be unimpeachable: the need to position public housing residents more in the mainstream rather than segregating them in high-rise public housing towers. Admitting that the latter were, as the authors state in the book's opening sentence, "a massive failure of public policy," they wondered whether the construction of partially subsidized mixed-income housing projects was the right solution.

In Chicago, the effort to rethink public housing began when Mayor Richard M. Daley launched the city's "Plan for Transformation" in 1999. The Plan had the goals of not only deconcentrating urban poverty through massive demolition and redevelopment—which involved relocating 56,000 people—but also achieving "cross-class integration." However, as Chaskin and Joseph (2015) argue, there has been a misunderstanding about community dynamics and how people actually relate to each other. Proximity alone does not achieve a meaningful type of social integration: people in a diverse setting find other, nonspatial ways of maintaining what sociologist Robert Park (1926) long ago called "social distance." The physical designs of mixed-income housing developments have not always helped, either—awkwardly placed units where the wealthy quickly retreat inside their homes, public spaces that are not accessible and badly designed, and services that never materialize.

DISCUSSION QUESTIONS

1. Do you think that deliberative, government-initiated social mixing is a band-aid approach that treats the symptoms rather than the cause of segregation and concentrated poverty? If you were in charge of housing policy, how would you do things differently?
2. Mixed-income housing sometimes fails because of the difficulty of finding consensus on the rules that should govern a development. Examples might be mandatory drug testing or limits on public gatherings. What method would you use to build consensus in a diverse, mixed-income housing development?

Activity 6.3: Mixing High Income and Low Income: Is It Necessary for Sustainability? (Debate)

Should governments adopt policies that work proactively to mix high- and low-income populations in residential areas? Some observers find that the assumptions underlying social mix goals are problematic: for example, that low-income residents benefit from high-income neighbors via access to social networks and job opportunities. Others question whether positive interaction and social support are in fact dependent on residential proximity. On the other side of the debate, some argue that the only way to ensure that there is an equitable distribution of the positive aspects of city life—services, amenities, transit, jobs—is to mix up the population. Such mixing counteracts social homogeneity with its exclusionary concentrations of wealth and its destabilizing concentrations of poverty.

The class will be divided into two teams to debate the following proposition: *Mixing high and low income in residential neighborhoods should be a key component of a city's social sustainability agenda.*

TOPIC 4: POLICIES THAT SUPPORT DIVERSITY

RESOURCES

Chapple, K., & Zuk, M. (2016). Forewarned: The use of neighborhood early warning systems for gentrification and displacement. *Cityscape, 18*(3). HUD USER. https://www.huduser.gov/portal/periodicals/cityscpe/vol18num3/article5.html.

Choi, M., Van Zandt, S., & Matarrita-Cascante, D. (2018). Can community land trusts slow gentrification? *Journal of Urban Affairs, 40*(3), 394–411.

In activity 6.2, students selected a neighborhood and measured its diversity. In this section, we'll think about how diverse neighborhoods might be strengthened from a policy and programmatic perspective. A key issue with socially diverse neighborhoods is that they always seem to be moving in and out of diversity—and, often, gentrifying. In other words, long-term stable diversity is difficult to hold on to. And yet, every city has at least *some* diverse neighborhoods. The question is how to sustain them. Diverse neighborhoods are not *always* gentrifying—some might be losing diversity due to disinvestment and decline. But in either case—losing diversity due to gentrification or losing diversity due to disinvestment—policies and programs are needed to help sustain diverse neighborhoods.

In a Western, capitalist democracy, it takes effort to sustain diversity. In the private housing sector, the market tends to procure and reward (from an investment standpoint) homogeneity, for reasons having to do with financial calculus. In addition, the rewards of homogeneity tend to be unevenly distributed, mostly benefiting white homeowners. On the public-sector side, federal, state, and local government policies have enforced racial segregation ranging from "redlining," in which people in African American communities were denied loans (see box 6.1), to subsidies for "white-only" housing developments (Rothstein, 2018).

A complication of sustaining neighborhood-level diversity is that such places require adequate services (schools, shops) in addition to a well-designed public realm—conditions that are desirable and that can eventually lead to loss of diversity. In fact, one of the most entrenched truisms about city design is that well-designed places quickly become unaffordable. It's a matter of short supply and high demand, coupled with the fact that affordability in desirable places goes against the basic laws of land use in the American housing market. The U.S. policy of using property taxes to fund schools and other amenities—instead of decoupling property taxes from neighborhood-scale investment—exacerbates the resulting inequities.

Since the market is unlikely to support a mix of unit affordability levels on its own, a wide range of policies and programs have been formulated over the years to help sustain diverse neighborhoods. Housing-related policies include the development of new mixed-income neighborhoods, inclusionary housing requirements, and condominium conversion ordinances. Local governments can also enact fair share housing programs, requiring new developments to provide a certain number of affordable housing units, which is likely to contribute to income diversity. Policies designed to limit gentrification may also have the effect of encouraging income diversity. These include rent control and strategies for subsidizing low-income housing in high- or middle-income neighborhoods.

The Community Land Trust (CLT) is a particularly interesting model for maintaining housing affordability in diverse neighborhoods. A CLT is a private, nonprofit

> **Box 6.4**
> Community Land Trusts
>
> A good source for information on CLTs is the nonprofit organization Grounded Solutions Network. Its mission is to cultivate lasting affordability in inclusive neighborhoods based on the idea that "equitable communities that are racially and culturally integrated are healthier and more sustainable for everyone." CLTs are often focused on homeownership but have also been used for rental housing, multiunit condominiums, live/work units, retail spaces, and homeless shelters, demonstrating that CLTs can be adapted to a variety of contexts and applications.

organization that owns land for the purpose of providing affordable housing. CLTs take land out of the speculative market, separating the cost of land from the cost of the housing unit and keeping it permanently affordable to low-income families. Long-term affordability is maintained through restrictions on resale. CLT land can be used as collateral to leverage new sources of funding for improvements that then go back into the neighborhood. The idea is to capture the "social increment" of land and development—an idea first advanced in 1879 by Henry George in *Progress and Poverty*. George proposed a single tax on land so that society—which had created land value in the first place—would reap the benefits.

Municipal policies can be aimed at ensuring that neighborhoods have a mix of housing types—different tenures (owner vs. renter occupied) and different forms and sizes, from single family to multifamily. Targeted preservation ordinances, density requirements, deconversion ordinances, and rent control might be all be used. A related and equally important goal is the mix of housing ages. In areas experiencing new construction, the retention of some existing housing stock, integrated and blended with new housing stock, would be an important strategy for maintaining diversity. Older units are often more affordable than new, which is why diverse neighborhoods tend to have historic layers.

Beyond measures to keep units affordable, which may only be applicable in gentrifying areas, other relevant policies involve making public investments—for example, to support new businesses or mixed-use housing developments. Because such investments can stimulate gentrification leading to displacement—especially in neighborhoods that have transit access, historic housing stock, pedestrian quality, and lots of potential for revitalization—there is a need to monitor policy effects. Some policies are relevant for neighborhoods losing diversity due to gentrification, while others are relevant for neighborhoods losing diversity due to disinvestment—policymakers need to understand and monitor the effects in either case.

And of course, prior to any kind of diversity-sustaining or diversity-enabling policy proposal, there would need to be resident support. Policymakers need to understand what kind of engagement process would be needed to build consensus and get resident buy-in for diversity-related policies.

> **Box 6.5**
> Diversity-Related Policies
>
> Two organizations that keep track of policies and programs that support diversity are PolicyLink and Community-Wealth.org.
> The following are examples of the kinds of policies these organizations advocate, which could be used to help sustain neighborhood diversity:
>
> - Policies to keep units affordable (community land trusts, inclusionary housing requirements)
> - Policies to entice the development of affordable units (tax credits, bonus densities)
> - Policies to retain rental units (condominium conversion ordinances)
> - Grants and loans to support small businesses, especially retailers, who service diverse neighborhoods

DISCUSSION QUESTIONS

1. What policies do you think are most important for sustaining neighborhood diversity?
2. If you were a city planner trying to support neighborhood diversity in a gentrifying area, and your proposal for rent control was strongly objected to by some neighborhood residents, what would you do?

Activity 6.4: Urban Displacement and Gentrification (Sequences)

Working in teams, students will create a sequence of events describing the trajectory of a neighborhood that has experienced gentrification and is now trying to overcome displacement. What led to the problem, and what can be done in the future?

Teams should consult resources on the *Urban Displacement Project* website, which was established to "understand the nature of gentrification, and displacement, and exclusion." The website tracks varying levels of gentrification pressure in cities around the globe. Teams might browse through the *Displacement Typology Maps* for more ideas about the process of gentrification. Teams should include steps that they think will help residents take more effective action against the problems of gentrification and displacement. Teams should present their sequences to the class, explain their rationale, and compare similarities and differences among other teams.

Activity 6.5: Community Land Trusts (Team Research)

Each team will be assigned to a community land trust (CLT), selected from the list of "model" CLTs maintained by Community-Wealth.org (*Community Land Trusts: Models and Best Practices*). Using their development examples, and using data from *Social Explorer*, try to determine if the CLT has successfully contributed to the creation of a diverse community by providing affordable housing in locations that would otherwise not be affordable. Does your team find evidence that the CLT model contributes to the creation of neighborhood social diversity? Discuss the viability of CLTs as a path to advancing neighborhood-scale diversity. What CLT seems to have had the most success in diversity terms?

TOPIC 5: ZONING AND DIVERSITY

RESOURCES

Brennan, M., Peiffer, E., & Burrowes, K. (2019, June 12). How zoning shapes our lives. *Housing Matters*. https://housingmatters.urban.org/articles/how-zoning-shapes-our-lives.

PBS NewsHour. (2021). How zoning can restrict, or even prevent, affordable housing. https://youtu.be/Joaprn70Zco.

Talen, E. (2012). Zoning and diversity in historical perspective. *Journal of Planning History*, *11*(4), 330–347.

Zoning regulates what kinds of land uses get developed and what kind of form development takes. Zoning codes thus have the power to influence multiple small decisions about urban growth and change—including whether that change leads to more, or less, diversity. Zoning impacts diversity because zoning can keep uses—and people, by way of regulated housing type—apart. The project "Desegregate Connecticut" (www.desegregatect.org) created a *Zoning Atlas* that shows in stark detail just how restrictive and homogenizing zoning can be (Bronin, 2021).

Organizations like the Form-Based Codes Institute (formbasedcodes.org) have been calling for the reform of zoning and the need to reverse the tendency for zoning to separate uses. Their argument is that diversity would be better tolerated if the form was attended to—good design helps smooth out the juxtapositions of uses and housing types that some people object to. Of course, reform efforts have to be weighed against local expectations, social traditions, and openness to change; neighborhoods will vary widely in their acceptance of social and economic diversity.

Zoning reform is about reversing the rules by which social segregation has been achieved: allowing multifamily units where they have been excluded and eliminating rigid building codes, minimum lot size, maximum density, minimum setbacks, and other ways of putting a cap on density, diversity, and infill. Other strategies for mixing housing include allowing corner duplexes, walk-up apartments on alleys, accessory dwelling units (ADUs), smaller lots, and duplexes, triplexes, and fourplexes that look like single-family homes (figure 6.4). Putting larger or more expensive housing in lower-income areas or restoring housing previously divided into smaller apartments are strategies that work in reverse: enabling higher-income housing in lower-income neighborhoods (strategies that would need to be monitored for gentrification effects).

DISCUSSION QUESTIONS

1. Do you agree or disagree that "form-based coding"—regulation of form rather than use—would be an effective way to garner support for diversity?
2. Some have proposed getting rid of zoning altogether and letting the market have free reign to build whatever the market dictates. Do you think this approach would help create more diversity within neighborhoods, or would it have the opposite effect?

NEIGHBORHOOD DIVERSITY

6.4 The Nettie Krouse Fourplex in Portland, Oregon. Finetooth, CC BY-SA 3.0 https://creativecommons.org/licenses/by-sa/3.0, via Wikimedia Commons. https://upload.wikimedia.org/wikipedia/commons/e/e4/Nettie_Krouse_Fourplex.jpg.

Activity 6.6: Suburban Homogeneity (Case Study)

Instructors will preselect socially homogeneous suburbs for teams to investigate. Teams will determine the causes, outcomes, and possible solutions to suburban homogeneity in preselected locations. Teams can use socialexplorer.com or census.gov to compile the facts of the case. What are the factors that contribute to social homogeneity? How much is it a problem, and for whom? What data can be gathered as evidence of causal factors? Teams will also investigate the impact of regulatory context: the zoning of the suburb. How much of the suburb allows only single-family housing? Where are duplexes, rowhouses, and apartment buildings permitted? Are these multifamily structures permitted as of right, or are there special requirements like variances needed? What changes to zoning law would be needed to encourage more diversity? What other social, cultural, and economic factors might need to change to grow diversity in this suburb?

LITERATURE CITED

Beatley, T., & Manning, K. (1997). *The ecology of place: Planning for environment, economy, and community*. Island Press.

Bishop, B. (2009). *The big sort: Why the clustering of like-minded America is tearing us apart*. Mariner Books.

Bronin, S. C. (2021). Zoning by a thousand cuts: The prevalence and nature of incremental regulatory constraints on housing. *Cornell Journal of Law and Public Policy*. Advance online publication.

Capra, F. (1996). *The web of life: A new scientific understanding of living systems*. Anchor Books.

Dramstad, W. E., Olson, J. D., & Forman, R. T. T. (1996). *Landscape ecology principles in landscape architecture and land-use planning*. Harvard University and Island Press.

Eckstein, N. A. (2006). Addressing wealth in Renaissance Florence some new soundings from the Catasto of 1427. *Journal of Urban History, 32*(5), 711–728.

Fischer, C. S. (1975). Toward a subcultural theory of urbanism. *American Journal of Sociology, 80*(6), 1319–1341.

Frank, L. D., Engelke, P. O., & Schmid, T. L. (2006). *Health and community design: The impact of the built environment on physical activity*. Island Press.

Freeman, L. (2006). *There goes the 'hood*. Temple University Press.

Galster, G. (2019). *Making our neighborhoods, making our selves*. University of Chicago Press.

Harris, B., Schmalz, D., Larson, L., Fernandez, M., & Griffin, S. (2020). Contested spaces: Intimate segregation and environmental gentrification on Chicago's 606 trail. *City & Community, 19*(4), 933–962.

Harvey, D. (2000). *Spaces of hope*. University of California Press.

Jacobs, J. (1961). *The death and life of great American cities*. Vintage.

Kempton, M. (1963). *America comes of middle age*. Little, Brown.

Lees, L. (2008). Gentrification and social mixing: Towards an inclusive urban renaissance? *Urban Studies, 45*(12), 2449–2470.

Lefebvre, H. (1991). *The production of space*. Blackwell Publishers.

Lesger, C., & Van Leeuwen, M. H. D. (2012). Residential segregation from the sixteenth to the nineteenth century: Evidence from the Netherlands. *The Journal of Interdisciplinary History, 42*(3), 333–369.

Mumford, L. (1938). *The culture of cities*. Secker & Warburg.

Mumford, L. (1968). *The urban prospect*. Harcourt Brace Jovanovich.

Nelson, Robert K., Winling, LaDale, Marciano, Richard, Connolly, Nathan, et al. (n.d.). Mapping Inequality. In R. K. Nelson & E. L. Ayers (Eds.), *American panorama*. Retrieved May 31, 2022, from https://dsl.richmond.edu/panorama/redlining/.

Owens, A. (2015). Assisted housing and income segregation among neighborhoods in U.S. metropolitan areas. *The ANNALS of the American Academy of Political and Social Science, 660*(1), 98–116.

Park, R. (1926). The urban community as a special pattern and a moral order. In E. E. Burgess (Ed.), *The urban community* (pp. 3–20). University of Chicago Press.

Reardon, S. F., Fox, L., & Townsend, J. (2015). Neighborhood income composition by household race and income, 1990–2009. *The ANNALS of the American Academy of Political and Social Science, 660*(1), 78–97.

Rothstein, R. (2018). *The color of law: A forgotten history of how our government segregated America*. Liveright.

Sharkey, P. (2013). *Stuck in Place: Urban Neighborhoods and the End of Progress toward Racial Equality*. Chicago: University Of Chicago Press.

Suttles, G. (1972). *The social construction of communities*. University of Chicago Press.

Talen, E. (2018). *Neighborhood*. Oxford University Press.

Talen, E., & Anselin, L. (2022). Tracking sixty years of income diversity within neighborhoods: The case of Chicago, 1950–2010. *Cities, 121*, 103479.

Talen, E., & Lee, S. (2018). *Design for social diversity*. Routledge.

Vale, L. J. (2013). *Purging the poorest: Public housing and the design politics of twice-cleared communities*. University of Chicago Press.

7

MOBILITY

Mobility—how people and goods move around and connect—presents an enormous sustainability challenge. Transportation networks (road and rail) and the trains, cars, bikes, and trucks that traverse them impact the environment, the economy, and social equity in significant ways. Cities have many mobility-related problems: congestion, safety, CO_2 emissions, and air quality, which in turn impact social life, economic growth, and public health. The inequity of access to transport options is a primary concern.

INTRODUCTION

RESOURCES

Holden, E., Gilpin, G., & Banister, D. (2019). Sustainable mobility at thirty. *Sustainability, 11*(7), 1965.

Singh, A., Hauge, J. B., Wiktorsson, M., & Upadhyay, U. (2022). Optimizing local and global objectives for sustainable mobility in urban areas. *Journal of Urban Mobility, 2*, 100012.

World Economic Forum. (n.d.). Why the future of sustainability will start with mobility. Retrieved January 20, 2022, from https://www.weforum.org/agenda/2021/04/future-of-transport-sustainable-development-goals/.

World Wide Fund for Nature. (2021). Sustainable mobility. https://wwf.panda.org/projects/one_planet_cities/sustainable_mobility/.

Mobility is the ability to move about the city. The EU Green Deal seeks to reduce CO_2 emissions by 55 percent by 2030 and aims for a 100 percent reduction by 2050. Any hope of accomplishing this must tackle the problems association with urban mobility. What this boils down to is that to support sustainable cities, mobility needs to be guided by two interrelated goals: reducing dependence on cars and making mobility more equitable. The first goal means that cities need to transition away from carbon-intensive modes of transport, whereby people can move around and access jobs, goods, and services by walking and biking, taking public transport, or accessing other forms of shared mobility (electric forms of mobility do reduce greenhouse gas emissions and are therefore part of the solution, but there are other associated problems, as we discuss below). The second goal means that the ability to move around the city and access jobs, goods, and services needs to be equitable and not just the province of wealthy, walkable urban neighborhoods.

TOPIC 1: THE TROUBLE WITH CARS

RESOURCES

Jacobs, J. (1961). Erosion of cities or attrition of automobiles. In *Death and life of great American cities* (pp. 338–371). Vintage.

Duany, A., Plater-Zyberk, E., & Speck, J. (2000). The American transportation mess. In *Suburban nation* (pp. 85–97). North Point Press.

Mobility suffers under the weight of car culture, which is certainly the reigning basis of mobility in the United States. As Jane Holtz Kay (1997) argued in *Asphalt Nation*, car culture obstructs mobility because of waste: "every attempt to move is fraught with wasted motion, wasted time, wasted surroundings, wasted money." A more specific issue is that the simultaneous accommodation of pedestrians and wheeled vehicles within cities is a constant problem. In fact, it has been for centuries—think of medieval cities with merchants selling their wares in the street while horse-drawn carts tried to maneuver through. Later, when roads turned to long straight avenues, the conflict between pedestrians and carriages speeding by was legendary.

> **Activity 7.1:** Mobility, Then and Now (Class Discussion)
>
> Watch the 1906 film of San Francisco, shot just before the 1906 earthquake, and then watch *Conquering Roads*. Think about the difference in mentality between urbanism then and now, as well as the role of cars in urban life being advocated. Think about how the form of the street and the behaviors observed (of people, vehicles, animals)—what strikes you as different from streets today, and do you think streets now are better or worse? What problem do you see with the "modern" point of view, and what do you think the creators of the 1937 vision got wrong? Which of the predictions espoused in this movie have come true, and which have not?

> **Activity 7.2:** The Right to the Street (Concept Map)
>
> Watch the 1972 Dutch film *De Pijp, Amsterdam* about how children demanded safe play space in a street in Amsterdam more than a generation ago. Working in teams, create a concept map of the many issues, concepts, topics, and ideas involved in this story of competing interests and ideals.

MOBILITY

7.1 Tony Garnier's *Cite Industrielle*, 1904. https://commons.wikimedia.org/wiki/File:Garnier-Tony,
_Cit%C3%A9_industrielle,_les_hauts_fournaux.jpg.

> Teams should start by discussing the wide array of social, political, economic, and technological processes and agendas involved in the quest to make streets appropriate for play space. Who are the different stakeholders involved, and what are their values and goals? What goals are shared, and what goals are in conflict? Identify possible coalitions, strong and weak alliances, and points of commonality and conflict. The primary idea or central concept in this case is the use of streets by children. From there, identify four to six closely related ideas or subordinate concepts (for example, a subordinate concept might be traffic calming or the rights of car drivers). Then determine a set of tertiary ideas, each connecting to one of the secondary ideas.

In the early twentieth century, modernists thought that the solution to the conflict between wheeled vehicles and pedestrians was to separate everything. Tony Garnier offered the *Cite Industrielle* in Paris in 1904, proposing a "machine-age community" of hydroelectric plants, aerodomes, and highways, all strictly segregated (figure 7.1). The proposal launched generations of planners and architects who envisioned a world of separation enabled by technological and engineering sophistication.

The problem with this vision is that the more separated things are, the less walking is possible and the more driving is required. Modernist ideas about urbanism that reached full flowering by the 1950s had the effect of increasing traffic on urban streets. In addition, accommodating cars on high-speed thoroughfares led to the mind-set that streets were no longer an important part of the public realm. The focus on designing for cars, speed, and unimpeded flow constituted a narrow conceptualization of settlement that discounted the complexity of cities and human behavior. Especially tragic has been the view that city streets should be widened and turned into fast-moving arterials to accommodate suburbanites fleeing the city—at the expense of inner-city neighborhoods.

Times have changed. Now, proclamations from the 1940s that adding more lanes or more highways solves traffic problems have been largely discredited (Speck, 2018). One reason is because of the "principle of triple convergence," a now recognized paradox that increasing road capacity is ultimately counterproductive because it induces more demand and lowers investment in public transport. This is why in Denmark, the policy is that when traffic gets too congested, officials close a lane of traffic! This forces people to adjust their behavior and take public transit. Of course, this only works if there is public transit to take, which, in many American cities, is not a convenient option.

Travel by individual, privately owned car can increase CO_2 emissions, but it also increases parking demand and traffic fatalities. Transportation accounts for about one-quarter of global CO_2 emissions and of that, about 75 percent comes from road vehicles (IEA, 2022). Finding more sustainable solutions to mobility can thus have a major impact on carbon reduction and air quality. In sum, the goal of sustainable mobility is to increase mobility without increasing emissions, congestion, and other harmful environmental effects.

During the pandemic, at first there was a decrease in world emissions of carbon dioxide. But now, according to the International Energy Agency, some worrying reversals of this trend are emerging: a decrease in transit ridership (perceived as unsafe) and an increase in private vehicles as a means of transport (IEA, 2022).

DISCUSSION QUESTIONS

1. Will Americans ever give up their preference for private automobiles, and should they have to? Public policies and investments in alternative transportation modes can only go so far. How do you get collective acceptance and have citizens change the way they think about mobility such that it is not entirely based on private automobiles? Will citizens make it their responsibility to reduce carbon emissions?
2. In response to efforts to curb private car use (through added taxes, for example), Americans sometimes argue that it is "their right" to drive if they so choose, and they should not be penalized for exercising this right. What might be an appropriate response to this argument?

TOPIC 2: WALKING AND BIKING

RESOURCES

Dumbaugh, E., & Gattis, J. L. (2005). Safe streets, livable streets. *Journal of the American Planning Association, 71*(3), 283–300.

Jacobs, A. B. (1995). Requirements for great streets. In *Great streets* (pp. 270–292). MIT Press.

Speck, J. (2018). *Walkable city rules: 101 steps to making better places*. Island Press.

By the latter decades of the twentieth century, many people shifted their focus away from the need for car-based access toward a concern for pedestrian life and walking, recognizing that walkability is the essential element that makes urbanism valued. Walkability is one of urbanism's greatest competitive advantage: the compensation for living in a compact urban place with minimal private outdoor space is that there are amenities, vitality, and street life. There has also been a strong linkage to public health, as streets that are pedestrian oriented are believed to have an effect not only on quality of place but on the degree to which people are willing to walk (Clemente et al. 2005). There is recognition, too, that an aging population needs walkable, bikeable cities (McCullough, 2020).

Many organizations have been created to advocate for walking, including working toward changing policies that foster a more walkable city. AmericaWalks.org maintains a list of walking organizations, and there are hundreds. PedBikeInfo.org is a comprehensive pedestrian and bicycle information center run by the University of North Carolina that has a large searchable collection of images, data, and training tools. Many of these strategies fall under the heading of "traffic calming" (Project for Public Spaces, n.d.). The concept is relatively recent, probably originating with the Dutch "woonerf" in the 1970s. Since then, traffic calming has been promoted as a way to cut vehicle speeds, reduce accidents involving cars and/or pedestrians, and lower noise levels and air pollution.

Can cities do away with cars altogether? In the United States, pedestrian-only "malls" that closed main streets to car traffic started popping up in the 1960s and 1970s, but most of them were later considered failures. According to one study (Matuke et al. 2021), surrounding population density and design factors—like creating a sense of enclosure with adequate building heights, having at least 50 percent window coverage, and including street trees, seating options, and appropriate lighting—are variables that can stretch the life span of a pedestrian mall. European cities seem to have more of these qualities than American cities; for example, Copenhagen's pedestrianized streets are legendary. Berlin's residents are using a grassroots approach and collecting signatures to create a car-free zone (Volksentscheid Berlin Autofrei, n.d.). In these kinds of cities, pedestrian streets work because merchants do not depend on parking close by—people are walking to stores and restaurants.

How do you get people to walk? Jeff Speck, author of *Walkable City* (2012) and *Walkable City Rules* (2018), has developed a "theory of walkability," arguing that to get people walking, there need to be four conditions: a reason to walk, a safe walk, a comfortable walk, and an interesting walk. Changing streets to be more walkable can be very cost-effective, like adding stop signs, eliminating center lines, and painting crosswalks and bike lanes. Nature in the city—street trees and green infrastructure—is also essential for enhancing slow modes of travel like walking and biking.

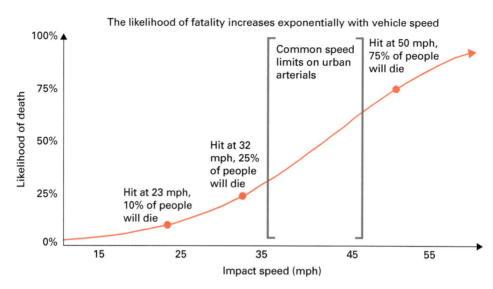

7.2 Vehicle speed and the likelihood of death. *Source*: National Association of City Transportation Officials. https://nacto.org/publication/city-limits/the-need/speed-kills/.

Box 7.1
The National Walkability Index

The National Walkability Index is a nationwide geographic data resource that ranks places (census block groups) according to their relative walkability. They have an Interactive Map to find the place where you live. It would be interesting to select locations in your city that, based on your local knowledge, are considered high-income and low-income areas and then find the data for percentage of zero-car households, percentage of one-car households, or percentage of two-plus-car households. Does anything in those statistics surprise you?

Box 7.2
Bicycle Networks

The organization PeopleForBikes has produced a Bicycle Network Analysis that uses Open-StreetMap (OSM) data (which is available worldwide) to measure how well bike networks "connect people with the places they want to go." The measure makes a distinction between "low-stress" and "high-stress" biking connections, based on street segment characteristics. It is especially useful for comparing the situation among different counties in terms of street pattern and high and low stress levels. PeopleForBikes also has a city rating system based on its bicycle network analysis and a community survey.

A primary focus of improving walkability is getting drivers to slow down. A rule of thumb (figure 7.2) is that a car traveling 35 miles an hour is seven times more likely to kill someone than a car going 25 miles an hour. This is why some cities (such as Tulsa, Oklahoma) are adopting a 25-mph speed limit throughout the downtown.

> **Activity 7.3:** Speed Limits and Car Crashes (Team Research)
>
> Explore New York City's excellent Crashmapper application. Explore different kinds of crashes, filtering by crash type and vehicle type. Find street segments that seem to have high crash events and investigate the on-the-ground conditions using Google streetview. Are there conditions that seem to characterize the higher crash injury locations? What are the posted speed limits in different crash locations, and do you see a patten to the relationship with crash variation? Do you see evidence that crashes are happening in places with wide streets, especially multilane one-way streets?

Besides reducing speed limits, there are some realistic changes cities can make to slow cars down and make things safer for pedestrians. Here are ten relatively low-cost interventions to make streets safer (adapted from *Optimizing Downtown Streets*, Mobile, Alabama, 2021).

1. There should be no more driving lanes than are needed (two lanes can handle 10,000 cars per day). In most cities, there is an oversupply of traffic lanes.
2. Lanes should be no wider than they need to be—the wider the lanes, the faster people will drive. Ten feet per lane is the standard for heavily trafficked commercial streets. Lanes do not need to be 12 feet. In residential streets, this width can be narrower.
3. Avoid multilane one-way streets. It encourages speeding and makes streets unsafe. Two-way streets are much safer.
4. Make street parking continuous; it protects the sidewalk.
5. Make a useful bike network. Bike infrastructure needs to get people where they need to go. It's important to include enough of bike lanes to reach a threshold such that people will actually start biking. There can also be different kinds of bike lanes, depending on traffic and street type. On busy streets, there needs to be protection between the parking lane and the traffic lane (called the "door zone"), and some don't.
6. Fewer center lines and more parking stripes—studies have shown that when a center line is removed from an arterial roads, people slow down. Parking and bike lanes should be well marked. If there is no centerline, people lose confidence that they can go fast; instead, they will drive more slowly and carefully.

7.3 A "neck-down" in New York City. *Source*: https://upload.wikimedia.org/wikipedia/commons/f/f7/Curb_extension_Bwy_%26_33_St_jeh.jpg. Attribution: Jim.henderson at English Wikipedia, public domain, via Wikimedia Commons.

7. Reduce or eliminate swooping geometries. Swooping streets cause people to drive faster. The goal is to tighten up the street to make it less swoopy, which will in turn make people drive slower. This might require reconstructing curbs, but it can also be accomplished using paint (called a "neck-down"; see figure 7.3).
8. Fewer sidewalk curb cuts—eliminate curb cuts where possible; they make pedestrians feel less safe.
9. No traffic signals that could be all-way stops. All-way stops are much safer, because people do not have the chance to speed through an intersection.
10. Make sure there are crosswalks where people are crossing.

DISCUSSION QUESTIONS

1. A wide variety of ideas are aimed at pedestrianizing the city, from street narrowing to street furniture. Where should interventions be prioritized, and based on what criteria?
2. If cars can't be eliminated altogether, can traffic at least be calmed? Traffic calming is the idea that thoroughfares should not be designed for cars to speed through town. Are there locations in your city where this has been tried, and has it been successful?

TOPIC 3: SHARED MOBILITY

RESOURCES

City of Buffalo, New York. (2020). The future of mobility: Remaking buffalo for the 21st century. https://www.cnu.org/sites/default/files/Buffalo_report-120220_spreads_sm.pdf.

Shaheen, S., Chan, N., Bansal, A., & Cohen, A. (2020). Sharing strategies: Carsharing, shared micromobility (bikesharing and scooter sharing), and innovative mobility modes. In *Transportation, land use, and environmental planning* (pp. 237–262). Elsevier.

Shearlaw, M. (2016, October 11). The Tallinn experiment: What happens when a city makes public transport free? *The Guardian.* https://www.theguardian.com/cities/2016/oct/11/tallinn-experiment-estonia-public-transport-free-cities.

Shared mobility broadly refers to the use of different types of transport (cars, mopeds, bikes, and scooters) through shared platforms, as well as trains and buses. The sharing of mobility has the potential to reduce CO_2 emissions, road congestion, and the need for parking spaces (thus reducing the need for asphalt, which in turn mitigates heat island effects and habitat loss). Shared mobility also contributes to equity because it provides affordable and sustainable mobility options for people at all income levels—especially those without access to privately owned cars. It is also important to note that there are many people who don't drive—children, many elderly, and people with disabilities.

Shared mobility is a kind of no-ownership model that rejects conventional production-consumption-disposal economics. Shared mobility creates a more efficient economy, one that keeps products circulating rather than disposing of them. It also significantly reduces the burden of ownership. Car ownership is a significant cost burden for low-income people.

What are the different types of shared mobility? Public transit—buses, trains, ferries—was the original shared mobility, and those types are still foundational. Shuttles, taxis, limos, and jitneys are longstanding forms of shared mobility. In the latter decades of the twentieth century, there was commute-based shared mobility, such as carpooling and vanpooling. Newer options include bikesharing, carsharing (Zipcars), ridesourcing (Uber and Lyft), ebike and scooter sharing, and microtransit—all of which are part of shared mobility. There are important linkages among them,

> **Box 7.3**
> Bus Rapid Transit
>
> Bus rapid transit or BRT is an alternative to fixed rail. Buses are given priority via a network of dedicated bus lanes so they can avoid traffic congestion. Such systems are considered more economical to build and operate. Curitiba, Brazil, is famous for having developed an innovative BRT system under the leadership of Mayor Jaime Lerner (established in 1974). The film *A Convenient Truth: Urban Solutions from Curitiba, Brazil* is a documentary about Curitiba's BRT system, for which the city was branded a "green city pioneer" with multiple international awards. Recently, ridership has recently declined, and the system has been criticized for failing to provide sufficient capacity.

too—for example, public transit creates an environment conducive to new forms of shared mobility, much more so than car-based urbanism. Also important to consider are the concepts of *intermodality* and *multimodality*, where two or more modes of transport are connected and users can move seamlessly between modes (Shaheen et al. 2020).

Shared mobility is now automated, radically altering urban economies via technological and digital innovation. This is good for urban sustainability because shared automated mobility relies, in part, on the sharing economy, reducing single-occupancy trips and car ownership. But it also puts a lot of strain on the urban curb zone. Curb zones are a scarce public resource, expected to accommodate food trucks, bus stops, e-scooters, bicycles, drones, vendor deliveries, ride hailing, and package delivery. The Open Mobility Foundation was formed by a group of U.S. mayors to produce "scalable mobility solutions"—tools and strategies for managing these dynamic and evolving curb zones (The Open Mobility Foundation, n.d.). The group argues that better management of curb zones is accomplished by better use of "curb data" and launched the Curb Data Specification (CDS) tool as a free, open-source, digital-first platform for cities to manage mobility in the curb zone.

All of these shared mobility data are creating privacy and cybersecurity concerns. The City of Buffalo produced a mobility plan calling for the creation of a "mobility and equity and innovation zone" where data would be used responsibly and safely "to improve efficiency in decision-making and promotion of public trust" (City of Buffalo, 2020). The downtown zone will be "a district that grows greener as it grows denser" by taking advantage of mobility innovations and compact living to achieve "opportunities that everyone can share" like net-zero emissions, parks instead of parking lots, and promoting walkability. The innovation zone will have mobility "spines" with protected lanes for escooters and bikes, smart traffic signals, and automated delivery options. Los Angeles has similarly proposed a "Transportation Technology Innovation Zone," where private-sector firms will have the ability to test out their new transportation technologies.

> **Activity 7.4:** Shared Mobility Data (Case Study)
>
> Estonia's capital Tallinn has a model that forecasts and analyzes the movement of pedestrians, cars, public transport, and trucks, distinguishing "12 different types of road users or user groups, including students, workers and pensioners" (Carey, 2022). The model also takes into account complicated movement patterns, like when someone drops off their child at school on their commute to work.
>
> Teams will investigate the issue of shared mobility by way of a case study analysis of Tallinn. Teams should consult the listed sources to find out more about Tallinn's mobility model.

One major benefit of shared mobility is its potential to vastly decrease the need for parking spaces. Car storage has been highly destructive to cities and urban neighborhoods, leading to poor design quality, acres of asphalt and impervious surfaces, and disruption of pedestrian life. Parking lots that front streets also have a way of lessening security. When people walk along sidewalks fronted by parking lots, they feel vulnerable, a problem Jane Jacobs wrote about extensively. This, in turn, compromises the economic health of commercial corridors, which, to stay viable, need residents and visitors to feel secure in them. It is of course necessary to address the interrelationship that exists between land use and transportation, because this ultimately affects the need for more parking.

Donald Shoup argued in his book *The High Cost of Free Parking* (2011) that free parking has led to a host of ills, from high energy use and environmental degradation to a damaged economy and financial strain. It is a self-propelling destruction, as more parking requires more driving, which in turn requires more parking. Researchers at MIT's Senseable City Lab have a project called "Unparking" demonstrating that on-demand mobility and self-driving cars have the ability to greatly reduce the need for parking lots (MIT Senseable City Lab, n.d.).

Converting parking lots to plazas is another option, where the plaza doubles as a parking area during certain times of the day. It becomes a multipurpose space, not simply a parking lot, an approach much more common in Europe than the United States. Ideally, the location would be near high-density residential areas so that as many residents as possible would be able to take advantage of the public space created (and not have to drive to it).

DISCUSSION QUESTIONS

1. Think about the tension between shared mobility innovation and privacy concerns. Do you worry whether data are being used safely and responsibly? Do you worry about being "tracked"?
2. In addition, do you wonder whether all this innovation in shared mobility is hurting rather than helping social and racial equity?

TOPIC 4: ELECTRIC CARS AND AUTONOMOUS VEHICLES (AVS)

RESOURCES

Henderson, J. (2020). EVs are not the answer: A mobility justice critique of electric vehicle transitions. *Annals of the American Association of Geographers, 110*(6), 1993–2010.

Madrigal, A. C. (2018, December 20). 7 Arguments against the autonomous-vehicle utopia. *The Atlantic*. https://www.theatlantic.com/technology/archive/2018/12/7-arguments-against-the-autonomous-vehicle-utopia/578638/.

Steuteville, R., & Crowther, B. (2020, May 27). Future of mobility in South Walton County. *Public Square*. https://www.cnu.org/publicsquare/2020/05/27/future-mobility-south-walton-county.

Much has been made of electric cars and automated vehicles as a solution to our CO_2 emissions problem. London's approach is a congestion charge if you drive within the congestion zone (which encompasses most of London), unless one's car meets ultra-low emission standards. London's mayor proposed replacing the congestion charge with a simpler system in which drivers pay per mile depending on how "green" their car is and whether they have access to transit. In other words, if one's car is polluting and one lives near transit, they are going to pay the cost. There would be discounts for low-income residents.

But electric vehicles have some significant limitations for addressing sustainability goals. Carfree Berlin makes the case against relying on electric cars as a solution to climate change with the following arguments zone (Volksentscheid Berlin Autofrei, n.d.). First, the production of electric cars involves a lot of raw materials, like copper, manganese, and aluminum, which are environmentally destructive industries that tend to be harmful to indigenous peoples. So rather than reducing energy and resource demand, electric vehicle (EV) production and consumption actually escalate it. Second, electric cars are still cars: they set in motion, on average, one ton of mass per person. That makes them dangerous. They do nothing to solve the problem of traffic-related injury and death, many of which involve pedestrians and bicyclists. Third, EVs are still cars that cause congestion and need roads and parking spaces. They are every bit as problematic in terms of land consumption as their gas-guzzling cousins, taking up space in the city that would be more sustainably resourced as green space or bicycling and bus lanes. Neither do EVs help with the promotion of compact, walkable neighborhoods. Another issue is that cars, no matter what propels them, create an enormous amount of dust from brakes and tires hitting the road. Finally, cars are high priced and unaffordable to many. The costs of owning and maintaining a car are a significant financial drain.

What about autonomous vehicles (AVs)? How do they relate to sustainable cities? Proponents, such as heavily invested car companies, argue that AVs will increase safety, greatly reduce parking demand, and reduce car ownership. Critics contend that AVs might be so popular that they will have the effect of increasing congestion and encouraging even more low-density sprawl. There is also the issue that AVs are

> **Box 7.4**
> The Car-Free City
>
> Culdesac, in Tempe, Arizona, is a car-free city with the motto "how you move defines how you live." It claims to be the first car-free neighborhood in the United States, or at least the first one intentionally built that way. The development is designed to be walkable, but the developers have partnered with mobility companies like Lyft to help people get around (residents receive a discount). E-scooters and bike parking are abundant.

likely to get hacked, since AVs will need to be programmed to stop for pedestrians, and people—think of a mischievous teenager—will start to feel empowered to stop them simply because they can. The surveillance and regulatory techniques necessary to determine and stop such behavior will be significant. But most of all, AVs will never have the capacity to move people that trains and buses have.

Activity 7.5: Are AVs Good for the Sustainable City? (Debate)

The class will be divided into two teams to debate the following proposition: *Automated vehicles are the basis of a more sustainable urban future.*

DISCUSSION QUESTIONS

1. In what ways do AVs have the potential to increase suburban sprawl and the spread of cities?
2. Car companies are putting a lot of stock in electric vehicles. Do you think the rise of electric vehicles is a game-changer for advancing sustainability goals?

TOPIC 5: EQUITABLE MOBILITY

RESOURCES

Barajas, J. M. (2021, October). Biking where black: Connecting transportation planning and infrastructure to disproportionate policing. *Transportation Research Part D: Transport and Environment, 99*, 103027.

Bierbaum, A. H., Karner, A., & Barajas, J. M. (2021). Toward mobility justice. *Journal of the American Planning Association, 87*(2), 197–210.

PeopleForBikes and Alliance for Biking and Walking. (n.d.). Building equity. Race, ethnicity, class, and protected bike lanes: An idea book for fairer cities. https://nacto.org/wp-content/uploads/2015/07/2015_PeopleForBikes-and-Alliance-for-Walking-Biking_Building-Equity.pdf.

Reducing carbon and congestion is not the only benefit of sustainable mobility—it is also the basis of a more equitable society. And yet, access to affordable, efficient transportation is an overlooked aspect of what it means for a city to be equitable and inclusive. Mobility is said to be inequitable because people in disadvantaged areas, often communities of color, tend to have very poor mobility options. Low-income neighborhoods often view transportation as a major obstacle to accessing what residents need in their daily lives—schools, stores, health care, and jobs. A more equitable system would provide good mobility options for low-income communities (which are often also communities of color). Many view mobility, like health care and housing, as a human right (Bierbaum et al. 2021). Employment professionals cite transportation as the number one barrier to work and employment, ahead of access to child care.

One approach to addressing transportation inequity is to reduce cost. Examples include making bus or train service free for the poor or subsidizing ride-sharing services. The country of Estonia has been giving residents free access to transit since 2012 and says that not only does it increase transit ridership, which is better for the planet, but it turns a profit since residents are required to live in the city and pay income tax in order to get the benefit. Some U.S. cities are instituting subsidies as well. The City of Oakland, California, has a plan to reduce car dependence and improve equity at the same time by giving low-income residents prepaid debit cards that they can use on sustainable modes of transport—public transit, bikeshare, and shared e-scooters.

The organization PeopleForBikes makes the case that bike infrastructure advances social equity (PeopleForBikes, n.d.). For one thing, biking is extremely cost-effective, making it an essential mobility resource for low-income people. In fact, poor people are more likely to rely on biking for transportation than other income groups (Cortright, 2019). A problem, however, is that there is an inequity of bike infrastructure—especially a lack of bike lanes in communities of color. In turn, this may be a factor in the unequal distribution of bicycle citations. The organization Biking where Black (Barajas, 2021) found that, in Chicago, tickets for bicycle citations in majority-Black census tracts were issued at eight times the rate as tickets in majority-white census tracts (and three times higher in majority-Latinx neighborhoods). One factor is that tickets are issued for riding on a sidewalk, which, in the absence of a bike lane, is often the only place bicyclists feel safe.

Box 7.5
Social Impact Calculator

The Center for Neighborhood Technology developed an eTOD Social Impact Calculator for Chicago that helps make the case for eTOD. Users can quantify the economic, social, and environmental benefits of eTOD projects. Users can type in an address and review the criteria used to calculate benefits under the tabs "Location Impacts," "Developer Resources," and "Demographic Trends." The purpose of the Calculator is to reveal "the benefits of affordable housing near transit to help developers select better sites, design unit mix, and educate the public and decision makers."

According to the organization America Walks, what is needed is a commitment to "mobility justice," defined as "a vision for a world rooted in social justice where people feel safe existing on the streets and can build lives experiencing the full joy of movement regardless of their race, religion, background, or physical ability." Mobility justice thus goes beyond "distributive justice," which only focuses on pedestrian upgrades, bike lanes, or making access to opportunity more equitable. Many "best practices" seem only to address "pavement, paint and place" rather than addressing discriminatory practices and committing to ensuring that placemaking efforts truly empower community members and give them an ownership stake. Instead, public funds are used to spur gentrification. Worse, "our own cultural resilience gets used against us to 'transform' urban neighborhoods from slums to arts districts" (Untokening, n.d.).

Equitable transit-oriented development, or eTOD, is another approach to mobility equity. It's about ensuring that there is sufficient affordable housing near transit. The environmental connection is that if there is too much distance between jobs and housing—referred to as "spatial mismatch"—it forces lower-income households to rely on cars, which not only is economically burdensome but also has the effect of creating more greenhouse gases.

Transit-oriented development (TOD), which is about creating compact neighborhoods around transit stops, typically a quarter or half mile, has gotten a lot of attention in the past few decades. If homes, jobs, stores, health care, and other amenities are anchored around a transit station, people can spend less time and money accessing what they need in their daily lives. *Equitable* transit-oriented development (eTOD) goes a step further and focuses on making sure that this access is available to anyone, regardless of income. Usually what this means is the development of subsidized affordable housing near transit.

DISCUSSION QUESTIONS

1. Do you agree that there has been too much focus on "pavement, paint, and place" in addressing mobility equity concerns?
2. Do you think transit-oriented development (TOD) is hurting or helping social equity goals?

TOPIC 6: MOBILITY AND URBAN FORM

RESOURCES

Barnett, J. (2003). Mobility: Parking, transit, & urban form. In J. Barnett (Ed.), *Redesigning cities: Principles, practice, implementation* (pp. 49–61). American Planning Association.

Farr, D. (2018). Mobility in walkable places. In D. Farr (Ed.), *Sustainable nation: Urban design patterns for the future* (pp. 236–257). Wiley.

Mobility encompasses more than modes of transport—it also concerns access to those modes and that has to do with land-use patterns and urban forms. What land-use patterns and urban forms make access more versus less equitable?

Mobility—getting from one place to another—is easier and less costly in terms of time and money if people and the goods and services they need are closer together and well connected (compactness often means density, a topic covered in chapter 5). In this section, we'll focus on how mobility is impacted by connectivity. How well people and places are connected is a central theme in sustainable urban design, as cities and neighborhoods that maximize connections between people and places are thought to be more vibrant and healthy (Salingaros, 1998).

Another important theme concerns the way in which connectivity is blocked. One prime example is construction of highways through neighborhoods, which severs social connection and often isolates people. This was a tragic occurrence of mid-twentieth-century urban renewal programs; however, people are now working to tear highways down and reconnect these severed neighborhoods.

Connectivity is accomplished via routes and thoroughfares, of which there are many different kinds. Figure 7.4 shows ten different thoroughfare types from the *Lexicon of New Urbanism* (Plater-Zyberk et al. 2014). The appropriate thoroughfare to connect one place to another will vary by scale and context. For example, urban designers may talk about regional connections in terms of highways and other major transportation routes, or about neighborhood-level connections via streets and greenways. Connections at smaller scales, such as by block, will involve even smaller types of routes and pathways. Connecting all types of spaces is important—public and private, residential and nonresidential, building and sidewalk.

A common strategy for promoting connectivity (and therefore mobility) is to ensure that streets are well connected. For example, regional-level connections are important since neighborhoods often benefit from being connected to the larger area or district they are a part of. However, streets have an obvious effect on separation and the disruption of neighborhoods. A recurring phenomenon in urban places is overly busy thoroughfares—streets with six lanes of traffic buzzing through the center of a residential area. Such roads have value as external connectors, but the cost is high—disruption of pedestrian quality and a lack of connectivity at a smaller scale.

Barriers and blockages in the city abound, disrupting connectivity and thus mobility. These may be direct—physical barriers like dead-end streets and cul-de-sacs—or indirect, like empty spaces, vacant lots, and parking lots, which can disrupt pedestrian routes. Mega-developments like hospitals and college campuses can act like clogged arteries—blocking connection and stranding residential areas around them. Barriers are also a product of streets that are too wide or unsafe to cross, such as wide arterials. There may be insufficient or even nonexistent crosswalks.

A focus on street connectivity draws attention to the size and shape of blocks, which have a significant impact on the corresponding patterns of movement. A gridded

MOBILITY

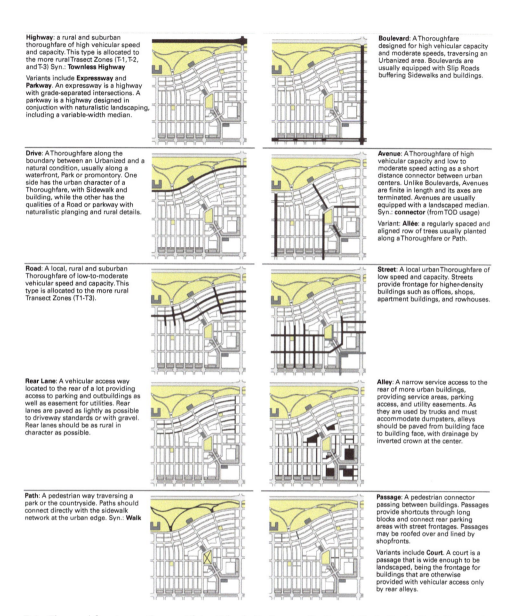

7.4 Thoroughfare types. *Source*: Plater-Zyberk, E., Longo, G., Hetzel, P. J., Davis, R., & Duany, A. (2014). *The lexicon of new urbanism*. Duany, Plater-Zyberk & Co.

street pattern is believed to offer the best connectivity because it provides multiple paths between points. This not only disperses traffic but also may allow pedestrians to navigate the shortest possible distance between two points. It is generally agreed that large-scale blocks, cul-de-sacs, and dendritic (tree-like) street systems are less likely to provide good connectivity.

Improving connectivity may involve adding, extending, or improving pedestrian paths, mid-block crossings, crosswalks, alleys, and bike paths. It may require developing additional, alternative routes to bypass the busiest streets or completing networks of greenways that are distinct (and separate) from automobile routes. It may involve ensuring that alleys are navigable, especially wherever streets don't provide a good route for pedestrians.

> **Activity 7.6:** Highway Teardowns (Sequences)
>
> Highways built in the 1950 and 1960s during urban renewal often barreled through viable neighborhoods, especially African American neighborhoods. In some cities, there is now an effort to tear these highways down, although some have argued that highway teardowns are the easy part (the class might want to read the article "Tearing Down Highways Won't Fix American Cities" by Gordon, 2021). As a class, students should first review the "Freeways without Futures 2021" report (CNU, 2021), which has case studies of fifteen highways that are "prime for transformation." Working in teams, select one of the fifteen highway case studies and create a sequence of events that describe (1) the key points in the history and current status of the highway and (2) the sequence of events that would need to occur to transform the highway. As a class, discuss what kind of impact would teardowns likely have and in what ways teardowns would advance equity goals.

Location efficiency is another important topic under the heading of mobility and the built environment. The goal is not unlike energy efficiency: to use resources in a smarter, more sustainable way. The Center for Neighborhood Technology, or CNT, developed the concept of location efficiency to inform homeowners about the costs associated with where they live. Since transportation costs are a household's second-highest expense after housing, residents who live in location-efficient neighborhoods will spend less time and money accessing their everyday life needs. And of course, they will also produce fewer greenhouse gas emissions. According to CNT, "Location-efficient communities are dense and vibrant, with walkable streets, access to transit, proximity to jobs, mixed land uses, and concentrations of retail and services" (CNT, n.d.).

Location efficiency goes beyond the usual measure of affordability—which is exclusively about mortgage and rent—to include the cost of transportation. This creates a different assessment of affordable living. Housing cost is supposed to be no more than 30 percent of household income, but by this measure, and based on housing cost alone, about 55 percent of U.S. neighborhoods are considered "affordable" for the average household. But when transportation costs are factored, the number of affordable neighborhoods drops to 26 percent. CNT has set a benchmark that housing and transportation costs combined should not exceed 45 percent of household income.

Transportation costs vary widely within and between regions. People who live in location-efficient neighborhoods—compact, mixed use, and with convenient access to jobs, services, transit, and amenities—tend to have lower transportation costs. On the other hand, location-efficient neighborhoods tend to have higher housing costs—specifically *because* of their efficiency. What is required, then, are affordable housing policies that address these transportation-related inequities.

The message of location efficiency is also relevant to employers, since where people work impacts daily commute. If workplaces are centrally located with good access to transit, with a variety of destinations nearby, employees do not have to rely on their personal vehicles for commuting and daytime trips. A location-efficient workplace results in lower congestion and pollution, as well as reduced cost burdens on local infrastructure. Indicators that measure location efficiency include worker commute mode-share, vehicle miles traveled (VMT), and workplace accessibility via transit.

The EPA and Smart Growth America teamed up to produce the Smart Location Calculator, which provides a way to estimate average employee commute time, VMT, and associated transportation impacts by workplace location. In this way, it is possible to compare the "location efficiency" of work sites. The goal is to improve facility site selection. Perhaps employers, supported by public policy incentives, can locate in a way that is locationally efficient.

Activity 7.7: Location Efficiency (Homework)

The Smart Location Calculator measures the environmental benefits of workplace location efficiency. The Smart Location Index (SLI) ranges in value from 0 to 100, where 0 indicates the least location-efficient site in the region, and 100 indicates the most location-efficient site. These scores are relative to the region. Enter your current address and discover your Smart Location Index (a score for your address of relative performance compared to the surrounding region) and your block group Smart Location Index (a score for the block group you live in). Then, click on "statistics" and record the vehicle miles traveled (VMT), emissions, access to transit, low-wage access, and mode split. Describe the kind of area you live in—is it car dependent, or are things within walking distance?

DISCUSSION QUESTIONS

1. Is it always a good idea to increase connectivity? Are there certain contexts where increasing connectivity might not be beneficial?
2. Do you agree with the goal of highway teardowns? Are there circumstances were tearing down highways in cities might not be the right approach?

LITERATURE CITED

America Walks. (n.d.). AmericaWalks.org.

Carey, C. (2022). Tallinn introduces predictive digital transport model. *Cities Today*. https://cities-today.com/tallinn-introduces-predictive-digital-transport-model/

Center for Neighborhood Technology (CNT). (n.d.). Location efficiency hub. https://cnt.org/projects/location-efficiency-hub.

City of Buffalo. (2020). The future of mobility: Remaking buffalo for the 21st century. https://www.cnu.org/sites/default/files/Buffalo_report-120220_spreads_sm.pdf.

Clemente, O., Ewing, R., Handy, S., & Brownson, R. (2005). *Measuring urban design qualities—An illustrated field manual*. Robert Wood Johnson Foundation. http:/www.activelivingresearch.org/downloads/fieldmanual_071605.pdf.

Congress for the New Urbanism (CNU). (2021). *Freeways without futures 2021*. https://www.cnu.org/highways-boulevards/freeways-without-futures/2021.

Cortright, Joe. (2019, May 28). Who bikes? *City Commentary*. https://cityobservatory.org/who-bikes/.

Gordon, A. (2021). Tearing down highways won't fix American cities. *Vice.com*. https://www.vice.com/en/article/4av3yn/tearing-down-highways-wont-fix-american-cities.

IEA. (2022). Global energy review: CO2 emissions in 2021. https://www.iea.org/reports/global-energy-review-co2-emissions-in-2021-2.

Matuke, S., Schmidt, S., & Li, W. (2021). The rise and fall of the American pedestrian mall. *Journal of Urbanism: International Research on Placemaking and Urban Sustainability*, 14(2), 129–144.

McCullough, K. K. (2020, March 5). Aging population needs walkable, bikeable cities. *Public Square*. https://www.cnu.orgpublicsquare/2020/03/05/aging-population-needs-walkable-bikeable-cities.

MIT Senseable City Lab. (n.d.). Unparking: MIT Senseable City Lab. Retrieved January 20, 2022, from http://senseable.mit.edu/unparking/.

Mobile, A. L. (2021). *Optimizing downtown streets final plan*. https://www.downtownmobile.org/uploads/pdf/MOBILEOptimizingDSReport.pdf.

PeopleForBikes. (n.d.). Peopleforbikes.org.

Plater-Zyberk, E., Longo, G., Hetzel, P. J., Davis, R., & Duany, A. (2014). *The lexicon of new urbanism*. Duany, Plater-Zyberk & Co.

Salingaros, N. A. (1998). Theory of the urban web. *Journal of Urban Design*, 3: 53-71.

Shoup, D. (2011). *The High Cost of Free Parking*. New York: Routledge.

Speck, J. (2012). *Walkable City: How Downtown Can Save America, One Step at a Time*. Farrar, Straus and Giroux.

The Open Mobility Foundation. (n.d.). https://www.openmobilityfoundation.org/.

The Project for Public Spaces. (n.d.). Traffic calming 101. https://www.pps.org/article/livememtraffic.

Untokening. (n.d.). http://www.untokening.org/updates/2018/1/27/untokening-mobility-beyond-pavement-paint-and-place.

Volksentscheid Berlin Autofrei. (n.d.). https://volksentscheid-berlin-autofrei.de/.

8

RESOURCE PLANNING IN CITIES

With nearly 4.5 billion people living on approximately 2 percent of global landmass, and projections of the world's urban population to increase from about 55 percent today to 70 percent by 2050, cities represent a great experiment in the allocation of scarce resources. While the compact city invites the sharing of resources and lower per-capita environmental footprints, cities on the whole are the home to three-quarters of the world's natural use resources, 78 percent of the world's use energy, and 70 percent of greenhouse gas emissions. Cities present many opportunities to be leaders in addressing climate change. Most cities have put forth ambitious targets for net zero, carbon-free energy, water conservation, and waste reduction. But many questions remain related city action on climate change. What strategies can be employed at a city level to address a global problem? What type of governance is needed but perhaps lacking? How is coordination created? Who is included in the decision-making, and how is success measured?

INTRODUCTION

RESOURCES

IPCC. (2022). Summary for policymakers. In H.-O. Pörtner, D. C. Roberts, E. S. Poloczanska, K. Mintenbeck, M. Tignor, A. Alegría, M. Craig, S. Langsdorf, S. Löschke, V. Möller, & A. Okem (Eds.), *Climate Change 2022: Impacts, adaptation, and vulnerability*. Contribution of Working Group II to the Sixth Assessment Report of the Intergovernmental Panel on Climate Change. https://www.ipcc.ch/report/ar6/wg2/

Cities are at the center of climate change planning and action. This is partly due to their status as the home to the majority of the world's population and as a result require large-scale consumption of natural resources and production of waste and pollution. Cities also face high vulnerability to climate damages as measured by impacts to human populations and infrastructure. Beyond the risks facing cities, there are important reasons why cities are well positioned to address climate change. Cities, as the largest population centers, are hubs of innovation and knowledge sharing with concentrations of capital, resources, and services. Cities are involved at least partly, if not fully, in the provision of the main resources of interest in climate planning—urban transport, energy, waste, and water. However, while cities possess many qualities suited for climate progress, action is constrained by governance and jurisdiction. Traditional city policies focus on buildings or site-level developments, zoning, transit planning, and local financing, all of which can influence climate planning and action but may not be enough for problems of a global scale (van der Heijden, 2021). However, addressing the climate crisis requires a rethinking of urban land use, as demonstrated throughout this book. This is inherently a task at the city level. As such, urban planners and policymakers seek to prioritize sustainable infrastructure development in the neighborhoods that most need it, decarbonize electricity, influence alternative modes of moving around the city, and engage both public and private actors to influence policy adoption at a larger scale (Tomer et al. 2021).

The majority of greenhouse gas emissions in cities come from the energy needed to power buildings and transportation, with the remainder from waste. In Houston, for example, buildings contribute 49 percent of greenhouse gas emissions, while transportation contributes 47 percent and waste with the remaining 4 percent (Green Houston, 2022). Cities also increasingly need to manage stormwater, water supplies, and urban heat to reduce vulnerability to climate change. Both the mitigation of greenhouse gas emissions from fossil fuels and the adaption needed to manage water and heat will require a rethinking of land use within the built environment. A holistic approach to climate planning engages the city at the individual scale of the citizen, the building scale to decarbonize, the neighborhood scale to improve land use, and the city scale to improve connectivity. This is accomplished through a range of strategies that motivate city denizens toward a low-carbon way of life.

The predicted climate impacts on cities depend on geography, topography, historical land use, climate, and adaptation strategies related to the built environment. Even cities in the Midwest, for example, which are relatively well protected from storm surges and wildfires, are projected to see extreme weather events leading to localized flooding and heatwaves. Disaster risk in cities is a function of both the exposure and vulnerability to natural and human-induced climate hazards. Human development and the built environment, as shown in figure 8.1, will have much to do with how people in cities will be affected by climate change.

DISCUSSION QUESTIONS

1. The IPCC report in the resources list identifies three categories of climate responses and adaptation options for urban and infrastructure systems: (1) green infrastructure and ecosystem services, (2) sustainable land use and urban planning, and (3) sustainable urban water management. What are some examples of city strategies

RESOURCE PLANNING IN CITIES

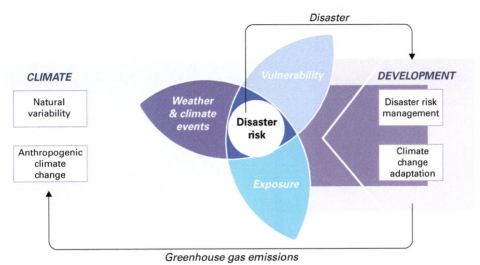

8.1 Disaster risk as a function of natural and human-induced weather and climate events (hazards), exposures, and vulnerability. *Source*: Lavell, A., Oppenheimer, M., Diop, C., Hess, J., Lempert, R., Li, J., Muir-Wood, R., & Myeong, S. (2012). Climate change: New dimensions in disaster risk, exposure, vulnerability, and resilience. In C. B. Field, V. Barros, T. F. Stocker, D. Qin, D. J. Dokken, K. L. Ebi, M. D. Mastrandrea, K. J. Mach, G.-K. Plattner, S. K. Allen, M. Tignor, & P.M. Midgley (Eds.), *Managing the risks of extreme events and disasters to advance climate change adaptation* (pp. 25–64). A Special Report of Working Groups I and II of the Intergovernmental Panel on Climate Change (IPCC). Cambridge University Press.

in each of these three categories? How do your examples influence each of the components of disaster risk: hazard/events, exposure, and vulnerability, if at all?

Activity 8.1: Greenhouse Gas Emissions Profile (Jigsaw)

Using your city of choice or your university campus (choose one with some available data), find the estimated proportion of greenhouse gas emissions from various sources (buildings, transportation, waste, other). Prepare a set of infographics using your software of choice or drawn by hand that describe the specific sources of greenhouse gas emissions. For example, you might want to note the total number of buildings and building footprint, describe the transportation profile of city or campus commuters, describe the sources of waste in detail, and document any recycling or conservation efforts. Students will compare and contrast profiles to see if cities or campuses face similar compositions or vary by region, population, politics, or other factors.

TOPIC 1: CITY STRATEGIES AND GOVERNANCE FOR CLIMATE ACTION

RESOURCES

Goh, K. (2020). Planning the green new deal: Climate justice and the politics of sites and scales. *Journal of the American Planning Association, 86*(2), 188–195.

Lin, B. B., Ossola, A., Alberti, M., Andersson, E., Bai, X., Dobbs, C., Elmqvist, T., Evans, K. L., Frantzeskaki, N., Fuller, R. A., Gaston, K. J., Haase, D., Jim, C., Konijnendijk, C., Nagendra, H., Niemelä, J., McPhearson, T., Moomaw, W. R., Parnell, S., Pataki, D., Ripple, W. J., & Tan, P. Y. (2021). Integrating solutions to adapt cities for climate change. *The Lancet Planetary Health, 5*(7), e479–e486.

O'Shaughnessy, E., Heeter, J., Keyser, D., Gagnon, P., & Aznar, A. (2016). *Estimating the national carbon abatement potential of city policies: A data-driven approach.* National Renewable Energy Laboratory, Task No. SA15.0803.

van der Heijden, J. (2021). When opportunity backfires: Exploring the implementation of urban climate governance alternatives in three major US cities. *Policy and Society, 40*(1), 116–135.

Integrated solutions for urban climate planning tend to focus on three-part strategies: *mitigation*—the reduction of greenhouse gas emissions targeted at the prevention of the changes in climate that lead to hazards, *adaptation*—changing land use and lifestyles for new climate-related conditions to reduce exposure, and *resilience*—investments to reduce vulnerability of the urban system through strategies designed to counter shocks and stresses.

Climate exposure and vulnerability vary across and within cities and relate to historic processes and social conditions. Therefore, adaptation strategies must be integrated across solution sets, climate targets, and equity goals. Solutions for urban climate planning come from a range of technological, nature-based, and socioeconomic strategies, but the integration across these sets is complicated by governance, scale, and, in some cases, ideology (Lin et al. 2021). Urban technological solutions include innovations in heating and cooling systems, advances in materials for buildings and streets, and smart system tools to automate responses to environmental conditions. Nature-based solutions align with the examples provided earlier in this book related to green infrastructure, while social solutions range from interventions to affect behavioral change to those intended to dismantle historical injustices that have led to chronic and persistent inequality. Clearly, no strategy is independent of others due to outcomes that affect both socioeconomic status and climate conditions. However, deliberate and intentional planning can be integrated across solution sets to devise sociotechnological solutions like active transit and shared mobility, technoecological solutions that employ smart systems to manage stormwater flows, and social-ecological strategies, which engage people in nature and green space activity (Lin et al. 2021).

Two challenges that persist in city-based planning for climate change and resource management relate to governance and spatial scales. Fragmented scopes, perspectives, and responsibilities in local governance affect conservation and adaptation policies for water, waste, and climate change (Koop et al. 2017), while cities that have set ambitious targets have failed to meet them due to a lack of power and capacity (van der Heijden, 2021). Additionally, while cities seek to achieve global targets for greenhouse gas emissions mitigation, misalignment of spatial scales of local governance and community participation with global activism for climate change adaption persist (Goh, 2020).

> **Box 8.1**
> Do City Climate Plans Meet Their Targets?
>
> Cities across the globe have set ambitious targets for greenhouse gas emission reductions, often in line with global aspirations and scientific projections. But how do we know if city efforts go on to meet their targets? Often plans are announced with great fanfare and increasingly include sophisticated dashboards to track progress. However, cities struggle due to the lack of capacity and jurisdiction (van der Heijden, 2021) and often fail to keep up with the data tracking or progress reporting, resulting in plans falling by the wayside. Further, when administrations change, new leaders want to announce and advance their own plan, disrupting continuity. Another potential concern is how greenhouse gas data are recorded and tracked in cities. While data collection for sustainability has improved greatly with the use of energy and water-tracking tools, as well as remote sensing and satellite imagery to document tree cover and green space, there still lacks accountability, capacity, validity, and verification of the methods. Cities self-report greenhouse gas inventories, which provide important information for mitigation target setting and assessment. However, these self-reported inventories have proven unreliable. A recent study of 48 U.S. cities found underreports of greenhouse gas emissions by 18.3 percent, on average resulting in questions about the reliability of city estimates (Gurney et al. 2021).

> **Activity 8.2:** Sustainable Plans and Sustainability Data (Homework)
>
> Revisit your submissions for activity 1.2 ("Sustainability Data and What It Reveals") and activity 1.3 ("Sustainability Plan Evaluation"). Choose one element of the sustainability plan that can be investigated using the sustainability data you found. How do the two align? What can this sustainability data tell us about the baseline (or benchmark) and progress made for this element of the sustainability plan? What would you recommend for data collection to be better aligned with this element of the sustainability plan?

Still, climate action at the city scale represents a major opportunity, and progress is being made. Cities have mobilized to set ambitious targets, often in response to stalemates at the country or global level. Many cities have relied on their ability to set mandates using traditional tools like building codes, waste and water regulations, and zoning laws but with a new climate framing. These "stick"-type approaches have been complemented more recently by incentive-based or "carrot" approaches, which, while more politically palatable, often have less impact due to the need for voluntary participation. Still, designed appropriately, a combined carrot–stick approach can induce more meaningful change and involve more involvement from a range of stakeholders to build consensus (van der Heijen, 2021).

As shown in table 8.1, cities possess a wide range of options for building, neighborhood, and city-scale climate action. While large-scale regional or national-level planning would facilitate and further city action in addition to cross-city collaboration and coordination, cities still have many tools at their disposal.

Table 8.1 Sample of city-level climate policy areas and possible strategies, adapted from O'Shaughnessy et al. (2016) and van der Heijen (2021).

City Policy Categories	Description	Examples
Building codes and reporting	Requirements for new construction and major renovations	LEED certification, energy and water efficiency standards, mandating benchmark reporting
Planning legislation	Zoning reform, smart growth planning, transit-oriented development, green infrastructure and open space preservation, recycling	Zoning based on ecological or environmental conditions (watersheds, coastal regulations), road diets, shared streets, green streets, green infrastructure incentives, coordinated waste, recycling and composting at the city scale
Public transit	Setting of fare structures, new transit services and infrastructure, high-occupancy vehicle prioritization	Improved reliability, safety, and frequency of transit; reduced fares; free rides; prioritization of transit to reduce commuting; congestion pricing; gas taxes
Building energy incentives	Reduce building energy use through incentives at the building or user scale, financing and rebates to support technology upgrades	Expediting permits for green building, green building challenges, rebates for energy-efficient upgrades and weatherization, green roof grant programs
Solar photovoltaic (PV) and distributed energy	Private and public deployment of rooftop or property-based distributed solar energy	Incentives and rebates, education, purchase agreements, facilitation of permitting and zoning codes, improve access
Municipal actions	Public building, transportation, water and greening requirements, better data tracking	Net-zero targets for city buildings, electrification of city fleets, residency requirements, greening city properties, tree planting, water rate setting

Sources: O'Shaughnessy, E., Heeter, J., Keyser, D., Gagnon, P., & Aznar, A. (2016). *Estimating the national carbon abatement potential of city policies: A data-driven approach.* National Renewable Energy Laboratory, Task No. SA15.0803.
van der Heijden, J. (2021). When opportunity backfires: Exploring the implementation of urban climate governance alternatives in three major US cities. *Policy and Society, 40*(1), 116–135.

The examples provided in table 8.1 put forth a range of options from mandated rules to incentive-based voluntary actions. To determine the appropriate set of strategies for comprehensive sustainability plans, cities rely on data for the carbon-abatement potential for mitigation strategies, projections of climate conditions for adaptation measures, and predictions of damage to improve resiliency and reduce vulnerability. In every case, in addition to the challenges outlined throughout this section, public participation is necessary to achieve ambitious, transformative, and meaningful climate change planning (Cattino & Reckien, 2021). The feasibility of implementing a program depends on the public's willingness to engage and if it meets their needs and improves their welfare. Further, a climate action will not be deemed successful overall if it exacerbates the unequal social conditions that have contributed to the disparities in climate exposure and vulnerability within cities. The new generation of sustainability plans has relied on participatory processes in the goal- and strategy-setting process and seeks to incorporate social equity as a guiding principle or primary evaluation metric. Chapter 9 will further explore the historical dimensions of disparities in environmental and health exposures and conditions and consider the validity and efficacy of efforts to undo social injustices and advance equity through climate planning.

DISCUSSION QUESTIONS

1. Table 8.1 puts forth a sample of city policy categories and strategies. What, in your opinion, is missing? Think of another area or strategy that a city could use and explain how it would advance sustainability.
2. Consider the governance capacity and environmental footprint of your college or university. What type of action do you think would create the most meaningful change related to climate change? You can define "meaningful" however makes the most sense to you (e.g., reduces the most greenhouse gas emissions, generates the most awareness, engages neighboring communities, advances city sustainability, attracts environmentally minded students and faculty, etc.). Explain how this action, which is within the jurisdiction of school decision makers, creates meaningful change, as you have defined it.
3. Why is scale so relevant to city and global climate action? What does this mean and how are scales aligned or misaligned at the city level?

Activity 8.3: City Strategies for Climate Change (Summary and Concept Map)

The previous section discusses various ways in which city governance structures can influence climate action. With a teammate, use one or more of the *city policy categories* from table 8.1 and the readings by O'Shaughnessy et al. (2016) and van der Heijen (2021) to design and *propose a detailed program* that advances climate action at the city level. Be specific and use the concepts outlined in this section—how does the program contribute to the *climate goals* of greenhouse gas *mitigation*, climate *adaptation*, or climate *resilience*? Does it address *hazards*, *exposure*, or *vulnerability*? Does the program rely on urban *technological*, *nature-based*, *social*, or an integrated *solution*? Write up a summary of your proposed program addressing these questions and then use a concept map to connect the policy category, the program, city actors, the climate goal, and the solutions.

TOPIC 2: BEHAVIOR AND RESOURCES

RESOURCES

Byerly, H., Balmford, A., Ferraro, P. J., Hammond Wagner, C., Palchak, E., Polasky, S., Ricketts, T. H., Schwartz, A. J., & Fisher, B. (2018). Nudging pro-environmental behavior: Evidence and opportunities. *Frontiers in Ecology and the Environment, 16*(3), 159–168.

Cities have been increasingly engaging its citizens in climate planning and actions. Participatory processes can better assess the needs of residents and create mutual understanding of economic, environmental, and social priorities within and across neighborhoods. While traditional city policy measures like updating building codes and more ecologically sensitive zoning laws are important for creating climate-sensitive development, carrot-type approaches based on voluntary adoption related to building energy use, transportation and public transit, and water and waste conservation are increasingly being explored. These incentive-based strategies, based on voluntary adoption, are often better received by the public, albeit perhaps less reliable and subject to changing conditions like economic downturns (or upturns) and COVID-19. These behavioral approaches, at times termed "nudges," are used widely in different forms and contexts and complement changes to the built environment to further motivate climate-friendly behavior.

Byerly et al. (2018) explore different types of interventions for six domains of pro-environmental behavior, as shown in table 8.2. While a broad definition of a nudge would include all of these types of interventions, a more specific view would only consider ones where the intervention is posed as a conscious and deliberate choice by the individual. Nevertheless, their framework evaluates traditional nudges, including education-based ones, which have been applied widely in energy, water, and waste conservation, and financial ones, which include rewards (e.g., rebates) or penalties (e.g., taxes, fines). They also evaluate research on contextual nudges, which are less direct and less explicit. These include attempts to change default behavior like automatic enrollment in energy conservation programs, commitments or pledges, priming through subconscious means, salient messaging, trusted messengers, and social norms such as peer comparisons (Byerly et al. 2018).

The research results showed that contextual nudges, including defaults and commitments, are most promising for reducing meat consumption and encourage waste reduction, while norms are effective for family planning and water conservation.

Table 8.2 Behavior-change adaptations that target decision-making in six domains where human behavior has large impacts on the environment.

City Policy Categories	Description	Examples
Building codes and reporting	Requirements for new construction and major renovations	LEED certification, energy and water efficiency standards, mandating benchmark reporting
Planning legislation	Zoning reform, smart growth planning, transit-oriented development, green infrastructure and open space preservation, recycling	Zoning based on ecological or environmental conditions (watersheds, coastal regulations), road diets, shared streets, green streets, green infrastructure incentives, coordinated waste, recycling and composting at the city scale
Public transit	Setting of fare structures, new transit services and infrastructure, high-occupancy vehicle prioritization	Improved reliability, safety, and frequency of transit; reduced fares; free rides; prioritization of transit to reduce commuting; congestion pricing; gas taxes
Building energy incentives	Reduce building energy use through incentives at the building or user scale, financing and rebates to support technology upgrades	Expediting permits for green building, green building challenges, rebates for energy-efficient upgrades and weatherization, green roof grant programs
Solar PV and distributed energy	Private and public deployment of rooftop or property-based distributed solar energy	Incentives and rebates, education, purchase agreements, facilitation of permitting and zoning codes, improve access
Municipal actions	Public building, transportation, water and greening requirements, better data tracking	Net-zero targets for city buildings, electrification of city fleets, residency requirements, greening city properties, tree planting, water rate setting

Source: Adapted from figure 3 in Byerly, H., Balmford, A., Ferraro, P. J., Hammond Wagner, C., Palchak, E., Polasky, S., Ricketts, T. H., Schwartz, A. J., & Fisher, B. (2018). Nudging pro-environmental behavior: Evidence and opportunities. *Frontiers in Ecology and the Environment*, *16*(3), 159–168.

Interestingly, they find traditional nudges using education or financial means were promising only for waste reduction. There were mixed results for transportation choices, with salience, commitments, and some financial intervention success but no effect from defaults norms or education (Byerly et al. 2018).

To further explore the use of nudge-type approaches on transportation choices, we can return to the built environment strategies to making streets safer in chapter 7. While street design can induce behavioral change, direct interventions may serve as a necessary complement and also contribute to addressing a wider range of driving-related externalities, including pollution. Kuss and Nicholas (2022) considered

Table 8.3 Potential of twelve interventions to reduce car use, adapted from table 7 of Kuss and Nicholas (2022), where + (yes), ~ (partly), x (no).

Intervention Type	Description	New	Suitable	Feasible	Potential to Reduce Car Use
App for sustainable mobility competition	Rewards for walking, cycling, or using public transit	+	+	+	High
Integrated car-sharing action plan	Car sharing between work and home	~	+	+	High
School travel planning	Advice for alternative transportation for students and parents	~	+	+	High
Workplace parking charge	Parking charges to fund public transport	+	+	~	High
Mobility services for commuters	Free transit passes, shuttles to work	~	~	+	Moderate
Parking and traffic control	Remove parking, replace with bike lanes, walkways	~	~	~	Moderate
Workplace travel planning	Remove parking, discounts for public transit, biking, advice for alternative transit	x	+	+	Low
Congestion charge	Drivers pay to enter city, revenue funds alternative transit	+	~	X	Low
Personalized travel planning	Discounted public transit, information	x	~	+	Low
University travel planning	Reduced parking, discounts for public transit, bikes, information	+	~	x	Low
Mobility services for university	Free public transit and shuttles	+	x	x	Low
Limited traffic zone	Car-free zones, fines fund public transit	~	~	x	Low

Source: Kuss, P., & Nicholas, K. A. (2022, August). Case studies on transport policy. A dozen effective interventions to reduce car use in European cities: Lessons learned from a meta-analysis and transition management. *Case Studies on Transport Policy.*

> **Box 8.2**
> The Power, Ethics, and Efficacy of Nudges
>
> While cities have policy tools and levers at their disposal, a question remains as to how directly people need to be invested in the idea of the sustainable city. Do people care enough about climate change to willingly change their lifestyles? Do they need to? The "nudge" approach popularized by Thaler and Sunstein (2009) relies on a premise they call "libertarian paternalism," which posits the notion that it is both possible and legitimate to affect behavior through interventions, while also respecting an individual's freedom of choice (Thaler & Sunstein, 2009). Even before the nudge terminology became widespread, neoclassical economists have relied on market-based approaches and incentives to shift the relative preferences toward low-cost or welfare-enhancing decisions. Further, by changing "defaults," proenvironmental behavior becomes the norm or status quo, without requiring a challenging transformation in cultural ways of life. This has powerful connotations in the environmental space, where advocates claim that if addressing climate change action continues to be politically polarizing, a series of nudges to make proenvironmental behavior the norm or default will make large-scale state mandated action less necessary. While the nudge approach has been increasingly favored and adopted for public policy, there are concerns about the ethical foundation of attempting to manipulate preferences (Shubert, 2017), the ability of nudges to crowd out larger climate efforts (Hagmann et al. 2019), and the neoliberal individualization of a collective action problem like climate change (Ciplet & Roberts, 2017). Can we reasonably expect billions of people to "do the right thing" when it comes to climate change or to be induced by a widespread system of nudges and incentives? Is the sum of many individual actions equivalent to collection action? While proponents of focusing on the individual may rely on a freedom of choice, others advocate for climate-affecting rules determined by the state. Ultimately, progress will be limited by a dichotomous framing of climate change solutions (privatization or state managed), and rather actions, policies, and behavioral change are needed at every scale (Ostrom, 2010).

twelve interventions with carrot-and-stick approaches to reduce car use in cities. The potential of these interventions to reduce driving was evaluated based on newness, suitability, and feasibility, as determined through expert interviews. Table 8.3, adapted from Kuss and Nicholas (2022), shows the interventions, which draw from a range of nudges, including financial, education, social norms, commitments, and messengers.

It is clear from their results that the suitability and feasibility of the interventions will affect their success. As such, built environment or infrastructural changes are necessary to align the suitability and feasibility of the strategies with the behavioral response.

DISCUSSION QUESTIONS

1. What is a nudge, and why is the approach so popular?
2. What are your concerns with using nudges to influence proenvironmental behavior?

3. Why are combinations of carrots and sticks, along with built environment or infrastructural changes, all likely part of the solutions for achieving climate-friendly cities?

Activity 8.4: Nudges for Campus Sustainability (Team Activity and Discussion)

Students will form teams to propose a nudge program for campus sustainability. Sample interventions include a pledge and competition, change of default, education program, peer comparisons, financial rewards, or others taken from table 8.2. State your team's goal and why it is important. For example, the goal may be to reduce food waste in dining halls; increase recycling across campus; conserve energy or water in dorms, labs, or recreational centers; encourage sustainable transit or walking; reduce plastic water bottle use; or something else that your team deems necessary and suitable for a nudge-type intervention. Write up the program, detailing why it is important, how the intervention would work, the expected outcomes, and possible challenges or obstacles. Also, note if any infrastructural or policy changes would be necessary to make this a success.

TOPIC 3: THE CIRCULAR CITY: WASTE

RESOURCES

Petit-Boix, A., & Leipold, S. (2018). Circular economy in cities: Reviewing how environmental research aligns with local practices. *Journal of Cleaner Production, 195*, 1270–1281.

The Circular Cities Declaration, a foundational principle of the European Green New Deal, defines a circular city as "one that promotes the transition from a linear to a circular economy in an integrated way across all its functions in collaboration with citizens, businesses and the research community" (Circular Cities Declaration, n.d.). More specifically, the U.S. Environmental Protection Agency defines a "circular economy" as a system that reduces material use, redesigns materials to be less resource intensive, and recaptures "waste" as a resource to manufacture new materials and products (U.S. EPA, n.d.). More simply, it is a transformation of a linear flow of resources to a circular one that minimizes waste in production, consumption, and disposal by taking into account the entire life cycle of products. Inspired by nature, where nothing is wasted, the circular flow attempts to close the loop between the extraction of raw materials and disposal of wastes.

This applies to all consumer products, including ones we consume daily, like food, clothing, cell phones, electricity, and drinking water to the building materials in our homes, schools, and offices. While recycling can improve circularity of resource flows, the hierarchy of waste minimization begins with reduction, then reuse and finally recycling. Recycling, while preferable to disposal, is a costly city service, suffers from low uptake in some cities due to myriad factors, and doesn't apply to all production or postconsumer waste. Plastic waste, believed by many to be recyclable, is in fact rarely recycled due to its material composition. As a result, plastic pollution is ubiquitous and contaminates all major ecosystems on the planet, with the United States being the largest contributor in the world (Law et al. 2020). Another increasingly noncircular flow of waste comes from electronics, particularly cell phones, which tend to be replaced at the end of contract life rather than the end of the actual phone's life span.

While cities have implemented large-scale recycling programs, experimented with incentive-based programs to divert waste from landfills, invested in food waste and composting infrastructure, and created drop-off facilities for pharmaceuticals, electronics, batteries, and other hard to dispose of products, opportunities for reduction and reuse of materials and products are often thwarted by the desire for the latest and newest products or lack of integration between those at the start and the end of a product's life cycle.

The circular economy scales up from a product level. Figure 8.2 from the Ellen MacArthur Foundation shows an industrial system that mimics the restoration activities of nature. The linear flow of materials to waste is still visible in the middle of the diagram, but the loops are closed for both finite materials and renewable resources to minimize waste along the way and at the postconsumer and postuser stages. It is a framework that can be adopted at a micro level for products like phones, clothing, and plastic bottles; a meso level for facilities, buildings, and eco-industrial parks; or a macro level for a city (e.g., a "net-zero" city).

The circular economy framework is certainly not without criticism or limitations. The feasibility of such business partnerships and alignment of flows will require significant investment and coordination. Further, by making waste an input to other processes, it becomes valuable, discouraging its prevention (reduction vs. reuse or recycling). Lastly, the circular economy emphasizes technology and free markets, creating a technocratic and economic ideology that provides uncertain contributions to a sustainable city (Corvellec et al. 2022).

Activity 8.5: Circular Flow (Concept Map)

Propose a circular economy approach or strategy for a type of resource (waste, water, energy, transport) or a product (e.g., food, plastic bags, clothing, smartphones, cars) in your city of choice. Draw a flow and show how your proposed approach would lead to conservation, reduction, or recycling. Present this to the class by showing the concept map and answer the following questions:

- Consider the integrated strategies discussed earlier in this chapter. Is your proposed circular flow a technological, nature-based, social, or integrated strategy of two or all three of these?
- How are people engaged in this framework? Does it create awareness? Behavioral change? Does it need to?
- Would this lead to meaningful change? Why or why not?
- What are the obstacles or barriers?
- What might work better?
- How is equity addressed in this framework? Are there equity concerns in implementing this idea? Is it fair?

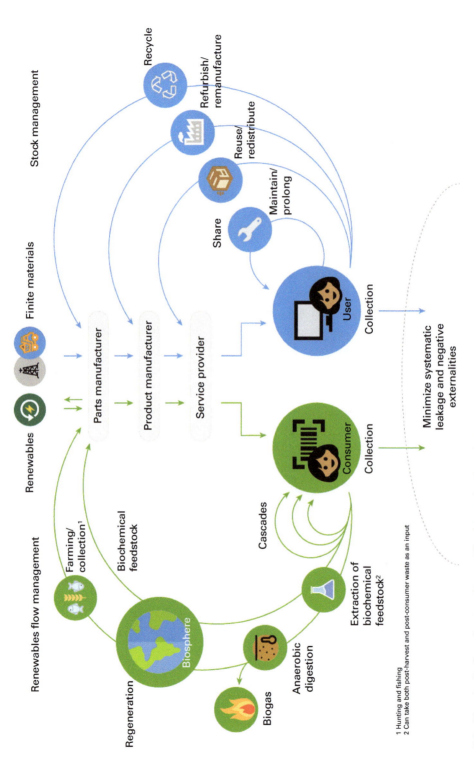

8.2 The circular flow diagram. *Source:* Ellen MacArthur Foundation.

TOPIC 4: THE NET-ZERO CITY: ENERGY

RESOURCES

Seto, K. C., Churkina, G., Hsu, A., Keller, M., Newman, P. W. G., Qin, B., & Ramaswami, A. (2021). From low- to net-zero carbon cities: The next global agenda. *Annual Review of Environment and Resources, 46*, 377–418.

Local governments and particularly mayors have focused the majority of climate efforts on the transition from conventional building stock to green buildings. This is logical since the operation of buildings contributes the most greenhouse gas emissions in cities and because of the municipal control over building codes and building energy incentives discussed earlier. There has also been significant market movement toward more energy-efficient buildings as a cost-saving tactic, as well as a way to draw higher rents. Green buildings therefore offer the most promising way for mayors to advance city sustainability plans as well as cross-city agreements around climate change (Lee & Koski, 2012). While LEED certification for new buildings has been around since the early 1990s, the certification has advanced to existing buildings, neighborhoods, and cities and communities.

The use of green buildings as a metric for sustainability has been criticized for the lack of oversight on the building's performance postcertification and in ongoing operations, as well as the concentration of green buildings in higher-income downtown core areas rather than prioritizing much-needed upgrades to deteriorating housing stock in lower-income and disinvested areas. Nevertheless, green buildings are a useful tool within the scope and capacity of city managers seeking to advance broad climate change goals (Lee & Koski, 2012).

Recently, the scaling up from buildings to cities has led to more than 800 cities globally committing to becoming "net-zero" cities as of December 2020 (Seto et al. 2021). While there is no formal or standardized definition of net zero, it refers to the condition of offsetting all CO_2 emissions within a building, city, or otherwise, with removal of carbon from the atmosphere. For cities, this is a long-term proposition, usually by somewhere between 2035 and 2050 from some baseline year set forth in their climate plan. Net zero then represents a type of circular flow, one that mimics nature or an ecological system where waste or leakages are minimized or eliminated. This is no small task. It requires a complete transformation of how cities are designed, built, and operated and must account for all flows in and out of the city's borders. Seto et al. (2021) detail how the idea of city transformation is not new by noting the well-known precursor movements to the net-zero city, including sanitary reform in the mid-1800s; the Garden City and the City Beautiful movement in the late 1800s, which embedded nature into city planning and brought grand designs of parks to cities, respectively; the eco-city and sustainable city of the 1990s, both of which considered ecology as a guiding principle in city design; the smart city approach, which originated in the early 2000s to create digitally connected cities; and the low-carbon city, which began to decarbonize cities on the path to the net-zero city (Seto et al. 2021).

> **Box 8.3**
> Nudges for Energy Conservation
>
> There has been increased adoption of nudge approaches to household energy conservation following research that began in the 1970s focused on reducing residential energy consumption and increasing energy efficiency (Byerly et al. 2018). Studies have shown that health information and social norms may be enough to induce energy conservation without any additional price incentives (Allcott & Kessler, 2019; Asensio & Delmas, 2015). This finding is important for framing energy conservation as individual responsibilities, but it is less clear if permanent changes to habits will result (Schubert, 2017). Further, it has been suggested that the interventions should be aimed at communities, engaging energy users as citizens, rather than as consumers (Heiskanen et al. 2010). It is also important to note that reducing energy use by any motivation will result in a cost savings. This is not necessarily the case with water conservation or waste reduction due to relatively low prices and flat-rate pricing structures instead of per-unit ones as more commonly used for residential energy. In these cases, conservation induced by nonprice mechanisms may not result in a perceived or actual cost savings, indicating that a combination of information and social norms, along with different pricing structures, may be necessary to nudge reductions of water or waste.

DISCUSSION QUESTIONS

1. How do buildings become "net zero"? Can cities truly achieve carbon neutrality? Is net zero or carbon neutrality a meaningful measure at the city scale? Why or why not?
2. Why do cities focus on green buildings in sustainability and climate change planning? Do you think this is the right approach for cities? Why or why not?
3. How do people engage in a net-zero building or a net-zero city? What combination of built environment and behavioral interventions is needed to advance carbon neutrality?

> **Activity 8.6:** Mapping Green Buildings (Team Research and Discussion)
>
> Using data on green buildings from a city of your choice or your campus, map the green buildings. What types of buildings are these? Are these all municipal buildings or privately owned buildings as well? Can you find anything about the performance of these buildings relative to nongreen buildings of the same type? Now, look into your city or campus sustainability or climate plan. What kind of role do green buildings play in meeting greenhouse gas emission targets? Do you think this is the way to achieve targets? What are some pros and cons of focusing on green buildings? Compare your map to a classmate who chose a different city or campus. What similarities and differences do you see? Is it hard to compare the two maps?

TOPIC 5: THE SPONGE CITY: WATER

RESOURCES

Fishman, C. (2011). Dolphins in the desert. In *The big thirst: The secret life and turbulent future of water* (pp. 51–87). Free Press.

Arnold, H. & Arnold, P. (2013). Pivot: Reconceiving water scarcity as design opportunity. *Boom: A Journal of California, 3*(3), 95–101.

The sponge city refers to a concept introduced in China as a model where rainwater is absorbed, stored, purified, and infiltrated back into the ground through a system of green infrastructure. Here, we consider it as a metaphor for a city, which, through low-impact development, green infrastructure, and holistic planning, utilizes blue (water-based), green, and gray spaces to maximize the conservation, reuse, and recycling of water while protecting both water quality and water supply. This closed-loop, circular flow of water is relevant in any city, regardless of the most pressing water issues.

Challenges to water management are ubiquitous, and no city or region in the United States is immune. In the American West, water stressors related to drought conditions and dwindling groundwater supplies continue to mount as cities seek technological solutions like desalination, nature-based strategies to restore water flows and capture scarce rainwater, and social and behavioral interventions like daily water use restrictions. In the Midwest and Northeast, aging storm and sewer water management infrastructure is strained by increases in rainfall, which is only projected to increase with climate change. In coastal areas, storm surges are becoming more frequent while rising sea levels will require a complete rethinking of city design and location. Access to affordable and safe drinking water is a pressing problem throughout the world, even in the United States, where nearly half a million households lack complete plumbing and 1,165 community water systems are in violation of the Safe Drinking Water Act with high burdens on the rural poor and indigenous peoples (Mueller & Gasteyer, 2021).

But opportunities exist. Awareness of water issues has increased dramatically due to high-profile cases and movements like the Flint, Michigan, lead poisoning disaster and Keystone XL and Dakota Access pipeline protests. Greater attention to oceans has arisen due to concerns about microplastic pollution, ocean acidification, and coral bleaching. In addition, water technology is advancing rapidly, and green infrastructure has become a design principle, particularly for stormwater management. The confluence of innovative water technology, awareness of water issues, and use of behavioral tools show promise for water management ahead.

Cities have traditionally managed both drinking water and storm and sewer water in conjunction with water management districts, private water utilities, and regional water agencies. For cities, water management is costly due to low prices, which do not cover the cost of water delivery and treatment, and the overuse of treated, potable water for nonpotable needs like toilet flushing, irrigation or lawn watering, and other outdoor uses. Water has historically been priced inefficiently low for all users,

Box 8.4
Stormwater Incentives

Stick approaches for stormwater management on private properties include low-impact development requirements and water-based building codes. Increasingly, cities are using a more flexible payments for ecosystem services approach, as described in chapter 4 for stormwater management. The economic reasoning behind such payment programs relies on the co-benefits to society taken on by private landowners by investing in green stormwater infrastructure. In Washington, D.C. and other cities, a stormwater retention credit trading program was developed to mimic other environmental markets that allow trading of environmental goals to meet targets at a lower cost. Developers who install green infrastructure on their property that manages stormwater above and beyond what is required received a "Stormwater Retention Credit" to sell back to the local government, thereby increasing the amount of green infrastructure investment on private land (DOEE, n.d.). Another common strategy in cities engages homeowners in on-site stormwater management by offering rebates and giveaways of rain barrels or native plants for rain gardens. A challenge for these programs has been the ongoing maintenance and operation of the green infrastructure, which can require specialized knowledge and attention to prevent overflows and increased flooding on site.

Box 8.5
A Day Zero for California?

In 2018, all eyes were on Cape Town, South Africa, as it approached "Day Zero," the day the municipal water network would shut down access for four million residents due to insufficient capacity in the freshwater dams. The prospect of running out of drinking water, combined with restrictions of only 13 gallons per person per day, left residents filling their water needs from natural springs, installing rain catchment systems, and even cutting down nonnative trees, which competed for water supplies. While some have questioned the fear tactics of the Day Zero narrative, Cape Town's water situation, a result of drought but mostly poor planning, management, and lack of investment in water infrastructure, remains dire and highly vulnerable.

As of 2022, Southern California is in the midst of a twenty-year drought, the worst in 1,200 years with little relief in sight. Ten months after a largely unsuccessful request by the Governor Gavin Newsom for Californians to reduce water use by 10 percent by their own accord, new restrictions went into effect in June 2022. These restrictions required water agencies to reduce water use by 20 percent, which in turn led to required reductions of up to 35 percent for some residential dweller. The restrictions equated to a per capita limit of 80 gallons per day, which means little water for outdoor uses such as car washing, lawn or garden watering, or pools.

even with graduated pricing by use or user. Raising the cost of water to encourage conservation is a challenging endeavor due to concerns about affordability and the commodification of water, and most demand-side management has focused on mandates related to building codes. Nudges or incentives, analogous to those related to stormwater, include rebates for improved water efficiency, financial rewards or peer comparisons for reduced household water use, and programs that encourage the adoption of native plants and other water-friendly alternatives to land and pavement. In dire situations, restrictions have been placed on the allocations of water to different users, resulting in strict limits on daily water use.

Advances in the design of water reuse and recycling are being made at the building and city scale. Supply-side interventions include expensive solutions like desalination of ocean water and more sustainable options like reuse of water for nonpotable purposes. Las Vegas is often touted as a model for its full reuse of indoor wastewater for outdoor purposes that few cities have emulated to date (Fishman, 2011). However, given its status as the driest city in the United States and its conspicuous use of water for golf courses and pools in a desert climate, there are still major concerns about water shortages, particularly as water supplied from the Colorado River continues toward historic lows.

A big concern for water lies in how existing disparities in access to safe drinking water, exposure to water pollution, and vulnerability to water disasters will be exacerbated from both climate change and climate change planning strategies. Some attention is being paid to these inequities as lead pipe replacement programs and green infrastructure investments are being prioritized in low-income, marginalized communities, while water shutoff programs are being reexamined following suspension during COVID-19 pandemic. Nevertheless, concerns continue to mount as water becomes scarcer and disasters become more frequent.

DISCUSSION QUESTIONS

1. Are sponge cities possible? Are incentives like the stormwater retention credit market or the rain barrel giveaway needed to make it work? Imagine how your city could be transformed to a sponge city. What would it accomplish? Is it feasible in your opinion? Why or why not?
2. What are some demand-side options to address water shortages? What are the supply-side options? What, in your opinion, are the best ways to deal with long-term water shortages?
3. Which sector uses the most water in the United States? Does this surprise you? Why or why not?

Activity 8.7: Understanding Your Water (Homework)

How much do you know about your water? Do you think about where it comes from when you turn on the tap? Do you think about where it goes when it disappears down the drain? What about when it rains? Where does the water go? Prepare an overview of your water using the guiding questions below.

- What is the source (body of water, groundwater) of your household water?
- Which agency is responsible for delivering it to your home (drinking water)? Which agency is responsible for treating it after it leaves your home (sewer water)? What happens to water outside your home after it rains (storm water)? Which agency is responsible for treating and managing it?
- How is the water priced (e.g., a flat rate for a given quantity, a per-unit charge, an increasing block rate—each block of water you use increases in price, or other) where you live? What is the cost per gallon? This may not be as straightforward as it seems. You might not know the specific rate, especially if you live in campus housing. If you don't have access to a water bill, try to find the average rate by user type in your town. There is likely a charge for water service and a charge for sewer services. List both, if possible.
- Find some estimates of water rates (prices) across the country or world. How does your water rate compare to this?

Activity 8.8: Your Water Consumption Story (Homework)

This section looked into demand-side water use restrictions in places with water shortage emergencies. Given you are a student likely living in a small unit without a big yard or private swimming pool, your water use is likely relatively small. However, do you know how much water you or your household uses? Do you know which activities use the most water? What is your indirect or "virtual" water consumption from your diet and lifestyle choices? In this assignment, you will conduct a ten- to fourteen-day water-tracking experiment. Split this time into two phases. In phase 1, you will assess your baseline by tracking water use through normal activity. Try not to do things differently, even though you might be inclined to do so because someone (you) is monitoring your water use! Then, in the second phase, engage in a conservation measure of your choice and keep tracking your water. How much did it affect your water use? A lot? A little? Do you think household water conservation can make a meaningful impact on water supplies? Prepare a report or story of your water-tracking experiment by compiling and visualizing your data and explaining your actions and outcomes. Here are the specific assignment details:

- In the first phase, record your baseline. How many showers do you take? How much laundry did you do? Did you use a dishwasher or wash by hand? Use the water calculators provided on the companion website to estimate how much water each activity uses or do a simple "flow test" at home. For example, see how much water you capture from a shower in thirty seconds and then use that to estimate your water use per minute spent showering.
- Compare your water use to a national or city average. Is it higher or lower? Why do you think this is?

- Also, consider your *"virtual" water footprint*. This is the amount of water required to make any items you consume, including food. Don't worry about sunk costs like the clothes you are wearing, but consider your behavior for the week. What is the virtual water from your consumption that week? Do some research on virtual water footprints and see the calculator on the companion website for guidance.
- Starting on the same day the following week, try your best to conserve your physical and virtual water use (don't overthink this or do any research, just do what is intuitive, and avoid doing anything that compromises the health of any living organism).
- Record all of your conservation activities. What are you doing that's different? Keep a log or diary.
- Make your best estimate at how much water you saved. To do this, you can explore the water calculators on the companion website or find others to consider both direct consumption of water and the indirect, virtual ones. You might also consider the co-benefits of your water saving (e.g., consider the carbon savings, energy savings).
- Report the results, and discuss your impact. Did anything surprise you? Is there anything you would change based on your results?
- What types of data, information, or calculations would you need to better inform your water practices?
- Recommend a behavioral intervention that would be likely to reduce residential water use. Explain why you think this would work based on your own experience, scholarly literature, and other research. Are there equity considerations of your intervention?

Summarize everything using a slide show, video presentation, or written report of three pages maximum, including any tables or visualizations. Include supplemental materials like a picture of your diary or log, an appendix on how you computed your baseline and made your estimates, and references or a bibliography.

LITERATURE CITED

Allcott, H., & Kessler, J. B. (2019). The welfare effects of nudges: A case study of energy use social comparisons. *American Economic Journal: Applied Economics, 11*(1), 236–276.

Arnold, H., & Arnold, P. (2013). Pivot: Reconceiving water scarcity as design opportunity. *Boom: A Journal of California, 3*(3), 95–101.

Asensio, O. I., & Delmas, M. A. (2015). Nonprice incentives and energy conservation. *Proceedings of the National Academy of Sciences of the United States of America, 112*(6), E510–E515.

Byerly, H., Balmford, A., Ferraro, P. J., Hammond Wagner, C., Palchak, E., Polasky, S., Ricketts, T. H., Schwartz, A. J., & Fisher, B. (2018). Nudging pro-environmental behavior: Evidence and opportunities. *Frontiers in Ecology and the Environment, 16*(3), 159–168.

Cardona, O. D., van Aalst, M.K., Birkmann, J., Fordham, M., McGregor, G., Perez, R., Pulwarty, R. S., Schipper, E. L. F., & Sinh, B. T. (2012). Determinants of risk: Exposure and vulnerability. In C. B. Field, V. Barros, T. F. Stocker, D. Qin, D. J. Dokken, K. L. Ebi, M. D. Mastrandrea, K. J. Mach, G.-K. Plattner, S. K. Allen, M. Tignor, & P. M. Midgley (Eds.), *Managing the risks of extreme events and disasters to advance climate change adaptation* (pp. 65–108). A Special Report of Working Groups I and II of the Intergovernmental Panel on Climate Change (IPCC). Cambridge University Press.

Cattino, M., & Reckien, D. (2021). Does public participation lead to more ambitious and transformative local climate change planning? *Current Opinion in Environmental Sustainability, 52*, 100–110.

Ciplet, D., & Roberts, J. T. (2017). Climate change and the transition to neoliberal environmental governance. *Global Environmental Change, 46*, 148–156.

Circular Cities Declaration. (n.d.). *Vision for a circular city*. https://circularcitiesdeclaration.eu/.

Corvellec, H., Stowell, A., & Johansson, N. (2022). Critiques of the circular economy. *Journal of Industrial Ecology, 26*, 421–432.

Department of Energy and Environment (DOEE), City of Washington, D.C. (n.d.). *Stormwater Retention Credit Trading Program*. https://doee.dc.gov/src.

Goh, K. (2020). Planning the green new deal: Climate justice and the politics of sites and scales. *Journal of the American Planning Association, 86*(2), 188–195.

Green Houston, Climate Action Plan. (n.d.). Retrieved June 11, 2022, from http://greenhoustontx.gov/climateactionplan/.

Grolleau, G., & McCann, L. M. J. (2012). Designing watershed programs to pay farmers for water quality services: Case studies of Munich and New York City. *Ecological Economics, 76*, 87–94.

Gurney, K. R., Liang, J., Roest, G., Song, Y., Mueller, K., & Lauvaux, T. (2021) Under-reporting of greenhouse gas emissions in U.S. cities. *Nature Communications, 12*, 553.

Hagmann, D., Ho, E. H., & Loewenstein, G. (2019). Nudging out support for a carbon tax. *Nature Climate Change, 9*(6), 484–489.

Heiskanen, E., Johnson, M., Robinson, S., Vadovics, E., & Saastamoinen, M. (2010). Low-carbon communities as a context for individual behavioural change. *Energy Policy, 38*(12), 7586–7595.

Koop, S. H. A., Koetsier, L., Doornhof, A., Reinstra, O., Van Leeuwen, C. J., Brouwer, S., Dieperink, C., & Driessen, P. P. J. (2017). Assessing the governance capacity of cities to address challenges of water, waste, and climate change. *Water Resources Management, 31*, 3427–3443.

Kuss, P., & Nicholas, K. A. (2022, August). Case studies on transport policy. A dozen effective interventions to reduce car use in European cities: Lessons learned from a meta-analysis and transition management. *Case Studies on Transport Policy*.

Law, K. L., Starr, N., Siegler, T. R., Jambeck, J. R., Mallos, N. J., & Leonard, G. H. (2020). The United States' contribution of plastic waste to land and ocean. *Science Advances, 6*(44), 1–8.

Lavell, A., Oppenheimer, M., Diop, C., Hess, J., Lempert, R., Li, J., Muir-Wood, R., & Myeong, S. (2012). Climate change: New dimensions in disaster risk, exposure, vulnerability, and resilience. In C. B. Field, V. Barros, T. F. Stocker, D. Qin, D. J. Dokken, K. L. Ebi, M. D. Mastrandrea, K. J. Mach, G.-K. Plattner, S. K. Allen, M. Tignor, & P. M. Midgley (Eds.), *Managing the risks of extreme events and disasters to advance climate change adaptation* (pp. 25–64). A Special Report of Working Groups I and II of the Intergovernmental Panel on Climate Change (IPCC). Cambridge University Press.

Lee, T., & Koski, C. (2012). Building green: Local political leadership addressing climate change. *Review of Policy Research, 29*(5), 605–625.

Lin, B. B., Ossola, A., Alberti, M., Andersson, E., Bai, X., Dobbs, C., Elmqvist, T., Evans, K. L., Frantzeskaki, N., Fuller, R. A., Gaston, K. J., Haase, D., Jim, C., Konijnendijk, C., Nagendra, H., Niemelä, J., McPhearson, T., Moomaw, W. R., Parnell, S., Pataki, D., Ripple, W. J., & Tan, P. Y. (2021). Integrating solutions to adapt cities for climate change. *The Lancet Planetary Health, 5*(7), e479–e486.

Mueller, J. T., & Gasteyer, S. (2021). The widespread and unjust drinking water and clean water crisis in the United States. *Nature Communications, 12*(1), 1–8.

O'Shaughnessy, E., Heeter, J., Keyser, D., Gagnon, P., & Aznar, A. (2016). *Estimating the national carbon abatement potential of city policies: A data-driven approach*. National Renewable Energy Laboratory, Task No. SA15.0803.

Ostrom, E. (2010). A multi-scale approach to coping with climate change and other collective action problems. *Solutions, 1,* 27–36.

Schubert, C. (2017). Green nudges: Do they work? Are they ethical? *Ecological Economics, 132,* 329–342.

Seto, K. C., Churkina, G., Hsu, A., Keller, M., Newman, P. W. G., Qin, B., & Ramaswami, A. (2021). From low- to net-zero carbon cities: The next global agenda. *Annual Review of Environment and Resources, 46,* 377–418.

Thaler, R., & Sunstein, C. (2009). *Nudge.* Penguin Books.

Tomer, A., Kane, J. W., Schuetz, J., & George, C. (2021, May 12). We can't beat the climate crisis without rethinking land use. https://www.brookings.edu/research/we-cant-beat-the-climate-crisis-without-rethinking-land-use/.

United States Environmental Protection Agency (U.S. EPA). (n.d.). *What is a circular economy?* https://www.epa.gov/recyclingstrategy/what-circular-economy.

van der Heijden, J. (2021). When opportunity backfires: Exploring the implementation of urban climate governance alternatives in three major US cities. *Policy and Society, 40*(1), 116–135.

9

ENVIRONMENTAL JUSTICE

Environmental justice has been posited as a framework, discourse, norm, value, rule, behavior, political position, and social movement. Ultimately, it is based on the principle that all people and communities have the right to involvement in environmental legal protections and outcomes. The achievement of environmental justice remains challenging as low-income individuals, people of color, and communities are persistently exposed to environmental toxins and increasingly to climate hazards. What are the historical causes of environmental injustices? How are social conditions and built environments precursors to environmental and health disparities? How can city climate action strategies and urban planning better address environmental injustices and ensure that social equity is at the forefront of all sustainability solutions? Are traditional theories, frameworks, and methods for designing cities, even sustainable cities, sufficiently critiqued to avoid exacerbating historical injustices? Are cities the appropriate scale and governance level at which to consider global environmental challenges? This chapter examines the concepts put forth in this book through the lens of environmental justice as the overarching goal for urban sustainability.

INTRODUCTION

RESOURCES

Lee, C., & Miller-Travis, V. (1987). *Toxic wastes and race in the United States*. United Church of Christ.

Mohai, P., Pellow, D., & Roberts, J. T. (2009). Environmental justice. *Annual Review of Environment and Resources, 34*, 405–430.

Aside from the similarities in organizing and activism, the modern environmental movement of the 1960s and 1970s (which evolved from many earlier movements) in the United States was largely decoupled from the civil rights movement of the 1950s and 1960s. In fact, while the discourse around environmental justice has been around for nearly forty years, it is only in recent years with the emergence of the Black Lives Matter movement and large-scale, often youth-led activism around climate change that the two pillars of social justice have converged into a more cohesive, albeit still not entirely unifying, movement. However, even as the environmental and civil rights movements of the twentieth century largely moved ahead down separate paths, the notion of environmental racism was concurrently the subject of ongoing protests by communities of color. The event that sparked national interest took place in 1982, when low-income African American residents of Warren County, North Carolina, took to the streets to protest a newly constructed hazardous waste landfill in their community. These protests, while not successful in preventing the site, are often cited as the launch of the organized movement for environmental justice. In response to the protests, the U.S. General Accounting Office (GAO) investigated the siting of four hazardous waste landfills in the southern United States. The report, released by the GAO in 1983, found that the communities surrounding the landfills were predominantly and disproportionately African American (Mohai et al. 2009), prompting a look into a nationwide problem. In 1987, the groundbreaking report *Toxic Wastes and Race in the United States* was published by the Commission for Racial Justice of the United Church of Christ (Lee, 1987). This report was the first to study the relationship between hazardous waste sites and the racial and socioeconomic composition of the surrounding communities. It found that race was the most significant predictor of hazardous waste facility locations and sparked an environmental justice praxis of social movements, public policy and planning, and scholarly research (Sze & London, 2008).

In the forty years since the Warren County protests, the question of environmental justice has gone on to be studied in depth by researchers from myriad disciplines and inclinations. Investigations have explored beyond the relationship of site location and racial composition of communities to examine disparities in pollution concentrations, tree cover, access to parks and urban green space, exposure to heat, flooding, energy systems, and localized contaminants and by-products like arsenic, lead, and other heavy metals. In addition, environmental justice as a principle has evolved to consider a broader definition of equity through community involvement in environmental decision-making and governance.

Today, environmental justice remains contentious in part because the documentation of a problem does not necessarily offer insight in how to solve it (Mohai et al. 2009). This is partly due to the relatively little scholarship on the causation and specific mechanisms by which environmental justice comes to exist (Sze & London, 2008). In practice, since there is rarely one person or entity responsible for cumulative exposures to health risks, there remains an absence of direct regulation, legal frameworks, and local laws through which to target environmental injustices. Further, dismantling the legacy of policies and built environment decisions that have contributed to segregation and inequality, and as a result environmental justice, is a complicated proposition.

DISCUSSION QUESTIONS

1. The reading by Mohai et al. (2009) provides several definitions and descriptions of environmental racism and environmental justice, as well as various explanations of why it exists. In your own words, describe the problem of environmental *in*justice, why it exists, and how it continues to persist.

> **Activity 9.1:** Contemporary Environmental Justice (Team Research)
>
> Do a quick Internet search to find an active or recent story related to environmental justice in or near your city. Since this will likely be a story of environmental injustice, what is the problem and what are the causes? Consider the direct cause (a chemical spill in a lower-income, minority neighborhood) and also the pathways that led to such an outcome (e.g., historical discriminatory zoning, unfair lending practices, lack of political representation). Who is involved? How is this being opposed and addressed? How is this situation related to the climate or sustainability plan of your city, if at all? Share your findings in class.

TOPIC 1: DEFINITIONS, ACTIONS, AND OUTCOMES

RESOURCES

Environmental Protection Agency (EPA). (2021). *Climate change and social vulnerability in the United States: A focus on six impacts.* EPA 430-R-21-003. www.epa.gov/cira/social-vulnerability-report.

Following the *Toxic Wastes and Race* report, environmental justice garnered attention at the federal level in the United States. In 1994, President Clinton signed the Executive Order 12898: Federal Actions to Address Environmental Justice in Minority Populations and Low-Income Populations (Exec. Order No. 12898). The order instructed federal agencies to "identify and address the disproportionately high and adverse human health or environmental effects of their actions on minority and low-income populations, to the greatest extent practicable and permitted by law." However, there are no clear or dedicated set of legal or litigation frameworks to enforce environmental justice, so the order lacks identification or enforcement capacity. The executive order was preceded by the establishment of the U.S. Office of Environmental Justice, created in 1992 to support the U.S. EPA in establishing community-based, public participatory programs, methods for data collection, and tools to help assess the environmental justice impacts of federal environmental regulatory actions. One such tool, the Environmental Justice Mapping and Screening Tool (EJScreen), seeks to provide the public with a nationally consistent set of data combining environmental and demographic indicators to support transparent decision-making around environmental justice (EPA).

In 2021, the EPA published *Climate Change and Social Vulnerability in the United States*, which focused on the greater exposures from climate changed faced by socially vulnerable groups, as defined by income, educational attainment level, race and ethnicity, and age. The study analyzed the disproportionate burden on these groups from climate change impacts related to air quality, coastal flooding, inland flooding, and extreme temperatures. The federal-level focus on the problem is necessary, but environmental justice is also a local problem. Segregation and inequality are heavily localized and embedded in local housing, education, transportation, and health systems.

Human influence on climate change has been well studied and given rise to the geological epoch known as the "Anthropocene." In addition to the human influence on the causes of climate change, design of the built environment and public policies affect exposures and vulnerability to climate change. What we consider to be natural hazards

---

Box 9.1
Redlining, Revisited

Historic programs that furthered inequality and segregation by race resulted in built neighborhood environments that continue to be sequestered, neglected, and underserved (Hendricks & Van Zandt, 2021). What necessarily follows this pattern of disinvestment are the conditions (less green space, fewer trees, more paved surfaces) that lead to urban heat island effects, leaving people and communities with greater vulnerability to environmental risk and climate hazards. One such program of institutionalized discrimination, called "redlining," was explored in chapter 6 of this book. Figure 9.1 from the *New York Times*, based on work by Hoffman et al. (2020), shows the current higher average satellite surface temperature in neighborhoods that were redlined in the 1930s. This aligns with a number of recent studies that have explored the relationships of redlining in cities across the United States to disparate environmental exposures, including air pollution (Lane et al. 2022), tree cover and ecosystem services values (Nowak et al. 2022), unhealthy food environments (Li & Yuan, 2022), siting of oil and gas wells (Gonzalez et al. 2022), flood risk (Linscott et al. 2022), and myriad negative health outcomes (Nardone et al. 2020).

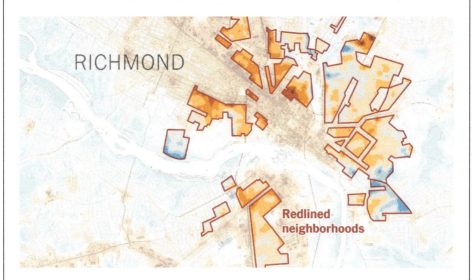

9.1 Historically redlined neighborhoods in Richmond, Virginia, show higher temperatures than compared to nonredlined areas. *Source: New York Times.* © [2020] The New York Times Company. All rights reserved. Used under license.

> **Box 9.2**
> Unnatural Disasters
>
> Heat waves are routinely cited as the leading weather-related cause of death in the United States, and concerns are mounting with projections of increasing heat waves under climate change. The greater exposure to heat combined with access to fewer resources and support institutions makes low-income people and communities of color more vulnerable.
>
> In *Heat Wave: A Social Autopsy of Disaster in Chicago*, a sweeping ethnographic account of the devastating 1995 Chicago heat wave, sociologist and New York University professor Eric Klinenberg investigated the social, political, and institutional factors that contributed to the death of nearly 700 people. His research demonstrated how social breakdown—the isolation of the elderly, abandonment of those in disinvested neighborhoods, and inaccessibility of public assistance programs—resulted in an unnatural disaster, one influenced by natural forces but dramatically shaped by human failure (Klinenberg, 2015).
>
> In addition to green infrastructure as a complement to gray physical, built infrastructure, social infrastructure is of critical importance in climate planning. As demonstrated by Klinenberg, it is the lack of social infrastructure, the networks that facilitate social connections and cohesion, that exacerbated the deadly damage of the Chicago heat wave.

(floods, earthquakes, storms) will result in a range of unnatural disasters that can best be mitigated through climate planning that examines historical, existing, and future built and social environments.

DISCUSSION QUESTIONS

1. Does your city or state have an office or officer of environmental justice? How is environmental justice addressed in your city or state, if at all?
2. Consider what you learned about redlining in chapter 6 and the resources in this section. How has institutionalized discrimination in housing and zoning led to environmental injustices? Can zoning reform address the problem of environmental injustice?
3. Can you think of a type of social infrastructure upon which you rely and depend? Explain.

Activity 9.2: Social Vulnerability and Climate Impacts (Team Research and Discussion)

Read through the EPA report *Climate Change and Social Vulnerability in the United States: A Focus on Six Impacts* from this section's resources. With your teammates, explain how the impacts in each category (air quality, coastal flooding, inland flooding, extreme temperatures) are analyzed and measured. For example, how does extreme temperature affect humans, especially those in what are defined by the EPA as the socially vulnerable groups? Answer this question for each of the climate impact categories in the report and comment on if you think these are the best ways to measure climate impacts on socially vulnerable populations.

Activity 9.3: Climate and Economic Justice Screening Tool (Jigsaw)

The Climate and Economic Justice Screening Tool (CEJST), created by the Council on Environmental Quality was created to identify communities that experience burdens in eight categories: climate change, energy, health, housing, legacy pollution, transportation, water and wastewater, and workforce development. Take a look at this tool and explore the maps in each of the categories. Create several maps of interest (either for different variables in one census region, or of the same variables in different census regions) and compare with your classmates. Do you see any patterns? Do you see high variability across census regions in the same city? What do these maps tell you? What additional information would you need to draw any conclusions?

TOPIC 2: ENVIRONMENTAL JUSTICE AND CLIMATE ACTION PLANNING

RESOURCES

Wolch, J. R., Byrne, J., & Newell, J. P. (2014). Urban green space, public health, and environmental justice: The challenge of making cities 'just green enough.' *Landscape and Urban Planning, 125*, 234–244.

The recognition of environmental pollution and climate change as a *social* problem has heightened the awareness that technology and financial investments on their own will not solve the climate crisis. Without attention on historical injustices and the discriminatory forces that shaped the built environment and social conditions, climate change is an intractable problem. To address the persistence of unjust environmental outcomes, the undoing of institutionalized discrimination and systemic racism must be at the forefront of climate policy. This recognition has taken shape in recent years, as evidenced by new strategies that place social equity as the forefront of climate planning. Most recently, developed city climate action plans emerge from a series of participatory processes with a focus on the neighborhood scale to account for the disparities that exist within cities.

Some cities have adopted the Green New Deal for cities, despite the lack of federal action, and there is growing interest and advocacy for reparations for climate change.

Box 9.3
Equity Considerations in Climate Planning

Cities have recently aligned their climate action plans with larger frameworks like the Sustainable Development Goals or Green New Deal. How do cities articulate their climate plans to equity? Is it a guiding principle? It is a consideration? Was there a public participation process, and what did it look like? How are the strategies aligned and integrated with other city priorities like housing, mobility, and schools? Given the established and demonstrated relationships between social conditions and environmental outcomes, the integration of such plans and policies is necessary for effective climate policy.

One question to consider is how the various urban sustainability topics covered in this book align with equity and environmental justice. For example, how do ecosystem services, green infrastructure, circular economies, and net-zero cities address equity? How might these as guiding principles exacerbate existing inequities? What about urban planning goals like density, neighborhood diversity, and shared mobility? Have traditional methods of urban planning historically created social conditions that lead to environmental injustices? To strive toward environmentally just cities, both the historical and contemporary approaches to urban planning, design, and sustainability must be critically assessed to understand the direct and indirect pathways to social and environmental inequities.

Activity 9.4: Climate Change Planning, Equity, and Environmental Justice (Team Research and Discussion)

The last section discussed the implications of using ecosystem services and green infrastructure in climate planning without specific attention to equity. What about the other aspects of urban sustainability discussed in this book? How do density, neighborhood diversity, mobility, the transect, circular economies, and net-zero goals connect to equity and environmental justice in cities? Students can be divided into teams to each consider one of the sustainability or planning goals and present it to the class.

TOPIC 3: URBAN PLANNING, SCALE, AND GOVERNANCE OF ENVIRONMENTAL PROBLEMS

RESOURCES

Meerow, S., Pajouhesh, P., & Miller, T. R. (2019). Social equity in urban resilience planning. *Local Environment, 24*(9), 793–808.

Urban planning has not traditionally been touted as a way to ensure environmental justice. This is in part due to scale but also due to the lack of attention in planning on broader social outcomes. In fact, some argue that failures in urban planning are the cause of certain environmental injustices like the Flint, Michigan, lead water crisis (Morckel, 2017). Others contend that conventional urban planning is not equipped to address environmental protection or social equity (Campbell, 1996).

Climate change is a global problem—one that cannot be addressed or solved by any actor in isolation. It requires international cooperation and recognition of the impact that a carbon-intensive economy has on the rest of the world. Climate justice, however, is broadly construed as a global environmental justice problem that manifests across space, time, race, and class. But, as the environmental justice movement and scholarship has shown, impacts can be heavily localized and disparate even across neighborhoods within a single city. Climate justice, therefore, considers the local impacts and experiences, disparate vulnerabilities, inequities, and the need for community engagement, sovereignty, and functioning (Scholsberg & Collins, 2014) and must be addressed at the urban scale (Bulkeley et al. 2014). Because climate change is a social issue, one influenced by social connections, built environments, and access to resources, it is inherently the most important problem for urban planners and policymakers to solve.

So, how might planners and policymakers take on a rethinking of traditional scales, approaches, and interventions? How might they consider local adaptation to a global problem? How can they deal with the misalignment of climate goals with equity?

In addition to putting social equity at the core of all urban environmental planning (Wilson et al. 2008), equity must be conceptualized beyond the distributional orientation to one with recognitional and procedural dimensions (Meerow et al. 2019). Instead of considering only the distributional effects of plans and policies, the recognition of historical discrimination and community needs, as well as the process by which goals are set, strategies are determined, and climate change is governed, must involve community participation (Meerow et al. 2019).

Returning to the climate disaster risk framework of hazard, exposure, and vulnerability, the greatest disparities lie in exposure, which can be addressed by climate change adaptation planning, including local land-use and zoning reform, and vulnerability, which requires attention to the social inequities created by historical discrimination, racism, and discriminatory public policy. A holistic solution for climate change ensures participation of all community residents in the planning process and carefully evaluates the outcomes and equity implications of all strategies. Because climate risks are disproportionate, it is important that climate solutions support the interests of those with the greatest need. Ultimately, the best chance at successful climate action looks not only to undo the fossil fuel–based structures upon which the world relies but also to dismantle the economics, policies, and social structures that have created historical inequities that lead to environmental injustices.

DISCUSSION QUESTIONS

1. Do you think that urban planners and policymakers can address environmental justice? What about climate justice? If so, how? If not, why not?
2. Do you think community participation in climate planning can ensure that climate strategies do not further disproportionate climate risks? Why or why not?
3. Is climate change a technological problem, a social problem, or both? Which is more pressing? Explain.

Activity 9.5: A Participatory Process for Climate Planning (Role-Play)

Students will be tasked with contributing to a climate action plan for their city. The city mayor has already decided in coordination with a neighborhood alderperson to invest in an innovative new design that converts parking lots into public green space in this underserved neighborhood. Students will be assigned to play the role of one of the following groups:

1. City climate action planners
2. City mayor and city council, including the alderperson (elected representative of the neighborhood)
3. Landscape architects/environmental designers
4. Resident of community (supporter—touts the health and social connection benefits of public green space)
5. Resident of community (opponent—the lack of transit access means we rely on cars and need parking. Also, public green space will lead to gentrification)

The role-play starts with a meeting of groups 1, 2, and 3. The three groups are discussing the task, which they have decided will provide great benefits to this underserved neighborhood. The plan was informed by the alderperson on the city council, a resident, and the elected representative of the neighborhood. After they meet, they decide on how residents will be engaged, if at all, and what the process will be. If students determine that any voices are missing from this process, they may create new groups and assignments. The game will stop once a consensus is reached on how to proceed with incorporating the intervention into the climate action plan, if at all.

LITERATURE CITED

Bulkeley, H., Edwards, G. A., & Fuller, S. (2014). Contesting climate justice in the city: Examining politics and practice in urban climate change experiments. *Global Environmental Change, 25*, 31–40.

Campbell, S. D. (1996). Green cities, growing cities, just cities? Urban planning and the contradictions of sustainable development. *Journal of the American Planning Association, 62*(3), 296–312.

Exec. Order No. 12898, 59, 32 Fed. Reg. (1994, February 16). https://www.archives.gov/files/federal-register/executive-orders/pdf/12898.pdf.

Gonzalez, D. J. X., Nardone, A., Nguyen, A. V., Morello-Frosch, R., & Casey, J. A. (2022). Historic redlining and the siting of oil and gas wells in the United States. *Journal of Exposure Science and Environmental Epidemiology*. Advance online publication.

Hendricks, M. D., & Van Zandt, S. (2021). Unequal protection revisited: Planning for environmental justice, hazard vulnerability, and critical infrastructure in communities of color. *Environmental Justice, 14*(2), 87–97.

Hoffman, J. S., Shandas, V., & Pendleton, N. (2020). The effects of historical housing policies on resident urban areas. *Climate, 8*(12), 1–15.

Klinenberg, E. (2015). *Heat wave: A social autopsy of disaster in Chicago*. University of Chicago Press.

Lane, H. M., Morello-Frosch, R., Marshall, J. D., & Apte, J. S. (2022). Historical redlining is associated with present-day air pollution disparities in US cities. *Environmental Science & Technology Letters, 9*(4), 345–350.

Lee, C., & Miller-Travis, V. (1987). *Toxic wastes and race in the United States*. United Church of Christ.

Li, M., & Yuan, F. (2022). Historical redlining and food environments: A study of 102 urban areas in the United States. *Health & Place, 75*, 102775.

Linscott, G., Rishworth, A., King, B., & Hiestand, M. P. (2022). Uneven experiences of urban flooding: examining the 2010 Nashville flood. *Natural Hazards*, *110*(1), 629–653.

Meerow, S., Pajouhesh, P., & Miller, T. R. (2019). Social equity in urban resilience planning. *Local Environment*, *24*(9), 793–808.

Morckel, V. (2017). Why the Flint, Michigan, USA water crisis is an urban planning failure. *Cities*, *62*, 23–27.

Mohai, P., Pellow, D., & Roberts, J. T. (2009). Environmental justice. *Annual Review of Environment and Resources*, *34*, 405–430.

Nardone, A., Chiang, J., & Corburn, J. (2020). Historic redlining and urban health today in US cities. *Environmental Justice*, *13*(4), 109–119.

Nowak, D. J., Ellis, A., & Greenfield, E. J. (2022). The disparity in tree cover and ecosystem service values among redlining classes in the United States. *Landscape and Urban Planning*, *221*, 104370.

Schlosberg, D., & Collins, L. B. (2014). From environmental to climate justice: Climate change and the discourse of environmental justice. *WIREs Climate Change*, *5*, 359–374.

Schubert, C. (2017). Exploring the (behavioural) political economy of nudging. *Journal of Institutional Economics*, *13*(3), 499–522.

Sze, J., & London, J. K. (2008). Environmental justice at the crossroads. *Sociology Compass*, *2*(4), 1331–1354.

United States Environmental Protection Agency (U.S. EPA). (n.d.). *EJ Screen: Environmental justice screening and mapping tool*. Retrieved June 11, 2022, from https://www.epa.gov/ejscreen.

Wilson, S., Hutson, M., & Mujahid, M. (2008). How planning and zoning contribute to inequitable development, neighborhood health, and environmental injustice. *Environmental Justice*, *1*(4), 211–216.

NOTES

INTRODUCTION

1. There are excellent books on sustainability and urban sustainability, which we draw from. For example, there is Roberston's (2017) *Sustainability Principles and Practices* and Brinkmann's (2021) *Introduction to Sustainability*, both of which step through a broad range of sustainability topics. Our task is to select from this voluminous sustainability material to make it applicable for an introductory, activity-based structure suitable for the classroom.

CHAPTER 3

1. Some of the methodology used in this exercise is adapted from the SmartCode Version 9 and Manual, available at http://smartcodecentral.com. Further information about the transect and the SmartCode is available at http://smartcodecentral.com.

INDEX

Adaptation, 116
Anthropocene, 140
Autonomous vehicles (AVS), 104–105
Avoided costs, 13

Biking, 96–101
Bus rapid transit (BRT), 102

Cabrini-Green (high- rise public housing project), 83
Capitalist democracy, 85
Car culture, 94–96
Central Park, 47, 49–51
Circular Cities Declaration, 124
Circular city, 124–125, 126
Classism, 76
Climate and Economic Justice Screening Tool (CEJST), 142
Climate change, 143
  human influence on, 140, 141
  and climate action planning, 142–143
  strategies and governance for climate action, 115–120
Community Land Trust (CLT), 85–86, 87
Cultural services, 11
Curb zones, 102

Density
  by design, 65–70
  locating, 63–64
  measuring, 64–65
  overview, 55–57
  preferences, 57–63
  walkscore *vs.*, 64
Discount rates, 14–15
Diversity, 73–78
  measuring, 80–82
  mixed-income housing, 82–84
  policies supporting, 84–87
  social mixing, 78–80
  zoning and, 88–89
Dominant density, 56

Economic welfare, 13
Ecosystem services
  assessment of, 17
  challenges to using, 22–23
  classification of, 12
  conceptual framework for, 10–12
  defined, 10
  measurement and methods for, 12–16
  monetary value of, 13
  overview, 9
  in planning and practice, 20–23
  urban ecosystem services, 16–19
Electric cars, 104–105
Environmentalism, 2–3
Environmental injustice, 22, 137, 139, 143
Environmental justice, 137–145
  overview, 137–139
  urban planning, 143–144

Environmental Justice Mapping and Screening
  Tool (EJScreen), 139
Equitable mobility, 105–107
Executive Order 12898, 139
Existence value, 13

Forgotten density, 56

General Urban Zone (T-4), 31
Greenhouse gas emissions, 114, 115
Green infrastructure, 22–23, 34, 39, 41–45, 97,
   118, 121, 129–131
  urban, 41–45
Green spaces, 37–51
  in cities, 37–38
  and human health, 38–40
  infrastructure, 41–46
  parks as, 47–51
  and planning scales, 44–45
Gross density, 65

Heat waves, 141
Hedonic pricing, 13
High-density cities, 56
Human habitats, 27–30
Human health, green spaces and, 38–40

Immersive environments, 26
Inequality, mapping, 77

Justice
  climate justice, 143–144
  distributive justice, 107
  environmental justice and, 142–143
  mobility justice, 107

Location efficiency, 110–111

Missing middle housing, 68
Mitigation, 116
Mixed-income housing, 82–84
Mobility, 93–111
  autonomous vehicles, 104–105
  car culture, 94–96
  electric cars, 104–105
  equitable, 105–107
  overview, 93–94
  shared, 101–103
  and urban form, 107–111
  walking and biking, 96–101
Monetization of nature, 16

National Walkability Index, 98
Natural Zone (T-1), 31
Nature, in city, 33–35, 37–38
Negative externalities, 20
Neighborhood diversity. *See* Diversity
Net density, 65
Net present value (NPV), 15
Net-zero city, 125, 127–128
Nonuse value, 13

Organisms, 28

Parks, as green spaces, 47–51
ParkScore®, 49
Physical environments, 28
Planning scales, green infrastructure and, 44–45
Positive externalities, 20
Principle of triple convergence, 95–96
Principles, sustainable city, 4–7
Provisioning services, 11
Public transit, 101

Racism, 76
Redlining, 77, 85,
  map, 77
  revisited, 140
Regulating services, 11
Replacement costs, 13
Resilience, 116
Resource planning, in cities, 113–133
  behavior and, 120–124
  circular city, 124–126
  net-zero city, 125, 127–128
  overview, 113–115
  sponge city, 129–133
Rural-to-urban transect, 25–35
  human habitats, 27–30
  nature in city, 33–35
  overview, 25–27
Rural Zone (T-2), 31

Shared mobility, 101–103
Simpson Diversity Index, 82
Social injustice, 56, 119

INDEX

151

Social mixing, 78–80
Space to Grow Green Schoolyards, Chicago, 18–19
Sponge city, 129–133
Sprawl, 25
Standard density, 64
Stated preference methods, 13
Suburban homogeneity, 89
Sub-Urban Zone (T-3), 31
Supporting services, 11
Sustainability, 3
  urban, 3–6
Sustainable city, 3–4
  density. *See* Density
  plan, 6–7
  values and principles, 4–7
Sustainable development, 3, 4
Sustainable urbanization, 4
Synoptic Survey, 35

Thoroughfares, 108–109
Transect, 25–35
  walk, 29
  zones, 30–33
Transit-oriented development (TOD), 107–109
  eTOD, 106–107
Travel cost method, 13

United Nations Development Programme (UNDP), 3
Urban Center Zone (T-5), 31
Urban Core Zone (T-6), 31
Urban green infrastructure (UGI), 41–45
  defined, 44
  densification, 44
  ownership, 44–45
  spatial scales, 44
Urban planning, 143–144

Valley section, 28
Values, sustainable city, 4–7

Walking, 96–101
Walkscore, 64
Weighted density, 64

Zoning, 81, 88–89